Biological Magnetic Resonance

Volume 33

More information about this series at http://www.springer.com/series/5693

Graeme Hanson • Lawrence Berliner
Editors

Future Directions in Metalloprotein and Metalloenzyme Research

Springer

Editors
Graeme Hanson (deceased)

Lawrence Berliner
Department of Chemistry and Biochemistry
University of Denver
Denver, CO, USA

ISSN 0192-6020
Biological Magnetic Resonance
ISBN 978-3-319-59098-1 ISBN 978-3-319-59100-1 (eBook)
DOI 10.1007/978-3-319-59100-1

Library of Congress Control Number: 2017947153

Printed on acid-free paper

This Springer imprint is published by Springer Nature
The registered company is Springer International Publishing AG
The registered company address is: Gewerbestrasse 11, 6330 Cham, Switzerland

Graeme R Hanson
(16 July 1955–25 February 2015)

To Lyn, Jeff, Harry, Johanna and Adrian

Preface

I have known Graeme Hanson for several decades. He and I planned this volume beginning in late 2014 and had it approved by Springer Science+Business Media for publication by the beginning of 2015. This book was spawned from our experience with two earlier volumes dating back to 2009 (28, *High Resolution EPR* and 29, *Metals in Biology*) that were very successful as based on chapter downloads from SpringerLink. Since then several new areas had emerged, and the demand was certainly there. Needless to say, Graeme's input was critical as he was one of the world's leaders in the general area of inorganic biochemistry/biology and magnetic resonance. We were in the process of signing the contract for this volume when he died. I decided to proceed with this volume, which carries his name as coeditor. Graeme and I also had applied and were granted a 1-month visiting researcher scheme for me at the University of Queensland (UQ) where we could both finalize and oversee the compilation of manuscripts for what was finally named *Future Directions in Metalloprotein and Metalloenzyme Research.* Fortunately, the visit still occurred, and I was able to spend a month at the Center for Advanced Imaging at UQ, a world-class institute that Graeme had a major role in establishing. I was honoured to present the first Graeme R Hanson memorial lectures at UQ, the first one on what would have been his 60th birthday. During my stay with his group, the planning of this volume was completed, authors were enlisted and their contributions now form the volume that follows this Preface. This book was completed in his memory while he is nonetheless a coeditor due to his enormous insight and contributions. What follows is my tribute to my dear friend, Graeme R Hanson.

Graeme and I would meet at various EPR meetings around the world and quickly became scientific colleagues and close friends. I had the pleasure of visiting him at the University of Queensland (UQ) during the early 1990s while on sabbatical in Melbourne. I was also honoured in 2008 to participate in the Sixth Asia-Pacific EPR/ESR Symposium (APES) in Cairns where Graeme was the organizer. Despite his being constantly busy and frequently unable to attend many sessions, he always had time for his friends. Graeme and I would schedule regular Skype calls to discuss science, the IES and issues of life and family. Graeme had the unique talent (like many excellent scientists) of taking on more responsibilities and projects that a normal scientist might not be able to juggle. He would ‘burn the midnight oil’ to complete manuscripts, grant proposals and the like; serve the profession; and, most importantly, never neglect his family obligations with intense interest in sharing their development and growth. Since Graeme had extensive experience as an officer in large international scientific professional organizations, such as the Society of Biological Inorganic Chemistry (SBIC), I would frequently request his advice on International EPR/ESR Society (IES) issues that I encountered as both Vice President–Americas and Interim President. As a result, we convinced Graeme to accept the position as IES Vice President–Asia starting 2015, a position he only was able to serve for less than 2 months.

The times I most cherish were his visits to Denver to attend the Rocky Mountain International EPR Symposia. Graeme, my wife Barbara and I would frequently attend an opera in Central City and have wonderful gourmet dinners together. The most memorable time was one late summer afternoon when Graeme went to Denver airport to discover his trip cancelled and rebooked for the next morning. I suggested that he find the airport bus that dropped him near our home and spend the night with us, and we spent a delightful evening dining, drinking and engaging in worldly discussions before dropping him at the airport the next day.

As mentioned earlier, we collaborated on several Biological Magnetic Resonance books. In collaboration with Takeji Takui, we completed another volume entitled: *Electron Spin Quantum Computing: Electron Spin-Qubit Based Quantum Computing and Quantum Information Processing* (2016). Consequently we jointly ‘produced’ four volumes together in this series, which we hope will remain as valuable reference sources for colleagues in the field.

I was also one of the few people outside of his close family circle with whom Graeme shared his cancer diagnosis. We discussed the treatment, the prognosis and his professional travel and scientific plans, which I kept totally confidential, until he informed the organizers that he was unable to attend the November 2014 APES/IES meeting together with the Japanese Society of Electron Spin Science and Technology (SEST). Although he suffered at least two setbacks and long hospital stays during the interim, I was shocked to hear that his weakened condition overcame him so soon. I think of Graeme every day; I remember our frequent telephone calls, past travels and always all our new plans for future scientific collaborations and publications. The profession has lost a real pioneer and contributor, and I (and my family) have lost a dear friend.

The chapters that follow are arranged in a sequential order from basic fundamentals of small molecules and model systems, including characterizing metal bridging in proteins and peptides (Simon Drew), copper(II) complexes of marine peptides (Peter Comba and Annika Eisenschmidt), high-spin Co(II) in model and metalloprotein systems (David Tierney and colleagues Robert Baum, Chris James), to enzymes such as the molybdenum-containing enzymes (Luisa Maia with Isabel and José Moura), CW and pulse EPR of cytochrome P450 enzymes (Jeff Harmer and colleagues Josh Harbort, James De Voss, Jeanette Stok and Stephen Bell) and finally the radical S-adenosylmethionine FeS family (Squire Booker with colleagues Alexey Silakov and Nicholas Lanz).

The authors are all world-class experts in their respective fields and devoted admirers of Graeme Hanson. This book is dedicated to his family: Lyn and children Jeff, Harry, Johanna and Adrian.

Denver, CO, USA Lawrence Berliner

Contents

An Isotopic Dilution Strategy for Characterisation of Paramagnetic Metal Bridging of Proteins and Peptides

Simon C. Drew

Abstract Electron paramagnetic resonance spectroscopy in conjunction with site-selective isotopic labelling provides a valuable tool for probing the local coordination environment of paramagnetic metal ions. Sometimes the metal ligands are inter-molecular in origin, such as during biological catalysis involving enzyme/metal-ion/substrate complexes or within oligomeric forms of peptides and proteins. Inter- and intra-molecular metal binding within peptide oligomers can be delineated by combining two fractions of peptide monomers containing the same amino acid sequence but having different isotopic content. Here, the principles of isotopic dilution in EPR spectroscopy are described and simulations provided to demonstrate its application to the identification of Cu^{2+}-bridging of peptide oligomers.

Keywords Copper • Isotopic labelling • Oligomer • Ternary • Superhyperfine

Introduction

Metal bridging interactions occur naturally as enzyme/metal-ion/substrate complexes during biological catalysis and as mixed-ligand (ternary) complexes in isolated chemical systems [1–3]. Metal bridging may also occur within oligomeric forms of peptides and proteins [4, 5]. When the metal ions are paramagnetic, electron paramagnetic resonance (EPR) spectroscopy can characterise the local coordination environment by resolving and interpreting metal-ligand hyperfine structure [6]. Although EPR does not readily discriminate between intra- and inter-molecular origins of the ligands, it is normally assumed in the case of metal-protein interactions that the ligands are intra-molecular in origin (derived from the same protein monomer). To distinguish these possibilities, one may replace the normally homogeneous peptide solution with a mixture comprising two fractions of peptide, each

S.C. Drew (✉)
The University of Melbourne, Parkville, VIC 3010, Australia
e-mail: sdrew@unimelb.edu.au

G. Hanson, L. Berliner (eds.), *Future Directions in Metalloprotein and Metalloenzyme Research*, Biological Magnetic Resonance 33,
DOI 10.1007/978-3-319-59100-1_1

containing the same amino acid sequence but having different isotopic content. For example, by combining peptides containing different $^{14}N/^{15}N$ content, a unique combination of ^{14}N and ^{15}N ligands becomes possible when intermolecular metal bridging occurs. As compared with intra-molecular coordination, this leads to a change in the metal-ligand hyperfine interactions that can be observed in the EPR spectrum [5, 7]. This procedure is conceptually similar to isotopic dilution methodology in NMR spectroscopy, which has been used to distinguish intra- and intermolecular contacts within peptide fibrils [8]. Application of this approach using EPR spectroscopy recently identified low molecular weight metal-bridged peptide oligomers [5, 7]. Here, the principles of isotopic dilution in EPR spectroscopy are described and simulations are provided to demonstrate its application to Cu^{2+}-bridged peptide oligomers.

Isotopic Labelling to Determine the First Coordination Sphere

To definitively identify metal bridging interactions, it is beneficial to first unambiguously characterise the coordination sphere of the metal ion. Continuous-wave (CW) EPR is capable of determining the type and number of ligands within the first coordination sphere. In the case of Cu^{2+}, assignments are sometimes based upon the empirical Blumberg-Peisach relations [9], although the use of these 'truth tables' is frequently ambiguous. This uncertainty is reduced when one can resolve the superhyperfine (shf) structure associated with the orientation-dependent hyperfine interactions of the unpaired electron with the metal nucleus and each of the ligand nuclei. The peak separations and intensities depend on the strength of the nuclear magnetic moment, the spin density at the nuclei and the orientation and distance of the ligand nuclei from the metal centre.

The identity of the metal ligands can be determined by introducing site-selective isotopic labels to specific peptide residues, then looking for variations in the shf pattern of the CW-EPR spectrum. For example, if the labelled residue provides a nitrogen ligand, then the multiplicity and spacing of the shf lines will change as compared with the unlabelled metal peptide complex due to the different nuclear spin and g factors of ^{14}N ($I = 1$, $g_n = 0.404$) and ^{15}N ($I = ½$, $g_n = -0.566$) [10]. In comparison with alternative approaches such as introducing point mutations in the peptide sequence, the isotopic labelling approach can be more sensitive to changes the ligand environment and preserves the native structure of the metal-protein complex. The utility of this approach for Cu^{2+} binding proteins and peptides has been demonstrated numerous times [5, 11–14].

Although pulsed EPR methods such as ENDOR and ESEEM spectroscopies have greater spectral resolution [15], the shf pattern within the CW-EPR spectrum enables the cooperativity of the ligands to be determined, because the number of shf lines follows a $\Pi_i(2I_i+1)$ rule (where I_i is the nuclear spin of the ith nucleus) [6, 16]. This multiplicative property is not generally observed for ESEEM and ENDOR, whose number of lines is additive with respect to the number of nuclei, unless peaks associated with double-quantum transitions can be resolved and properly quantified.

Isotopic Dilution to Identify Metal-Bridged Peptide Oligomers

The isotopic labelling technique described in the previous section provides information about the local coordination environment of the metal ion. It can be further exploited to provide information pertaining to the quaternary structure in instances where the metal ion resides in a bridging position between two identical or distinct peptides. To determine the extent of intermolecular bridging, one may study a solution of with some peptide monomers having some or all of their natural abundance isotopes replaced with other stable isotopes. The probability $P_n(j,k)$ of forming an nth order closed-chain oligomer $M_nA_jB_k$ in a solution containing a mixture of two peptides (denoted 'A' and 'B') with different isotopic composition is given by the binomial expression [7]:

$$P_n(j,k) = \binom{n}{j,k} (p_A)^j (p_B)^k \tag{1}$$

where the binomial coefficient accounts for the fact that the statistical probability of forming $M_nA_jB_k$ is $\binom{n}{j,k} = \frac{n!}{j!k!}$ times more likely than forming M_nA_n and M_nB_n. Thus, for the antiparallel dimers shown in Fig. 1, there are $\frac{2!}{1!1!} = 2$ equallyprobable Cu_2AB structures (Figs. 2 and 3). Provided the isotopic labels are introduced to metal ligands, then a solution containing a mixture of A and B introduces four geometrically equivalent, yet magnetically distinguishable, first coordination spheres of which two are unique to intermolecular metal bridging. For an equal concentration of A and B, one has $p_A=p_B=½$ and each magnetically distinguishable coordination sphere occurs with equal probability of 0.25. This is illustrated for two specific examples in Figs. 2 and 3, while Table 1 and Table 2 provide additional examples than can be verified in a similar manner. Analysis of an open-chain metal-bridged oligomer $M_mA_jB_k$ ($j + k = n$, $m = n - 1$), where all 2^n distinguishable permutations occur with equal probability of $(½)^n$ when [A] = [B], again yields an equal probability of 0.25 for each of the four distinguishable coordination spheres [5, 7].

Methods

Experimental Preparation

To implement the isotopic dilution method, at least one ligand from each subunit of a bridged oligomer must contain an isotopic label in order to distinguish the EPR spectra of monomer and metal-bridged oligomers. Thus, a minimum of two peptides (A and B) containing different isotopic labelling patterns should be synthesised, leading to the preparation of at least three different samples for EPR analysis: (1) M/A 1:1, (2) M/B 1:1, (3) M/A/B 1:½:½. Although the shf lines are frequently hidden beneath the inhomogeneous linewidth, resolution can be improved by working at low microwave frequencies to reduce '*g* strain' broadening [17, 18]. For

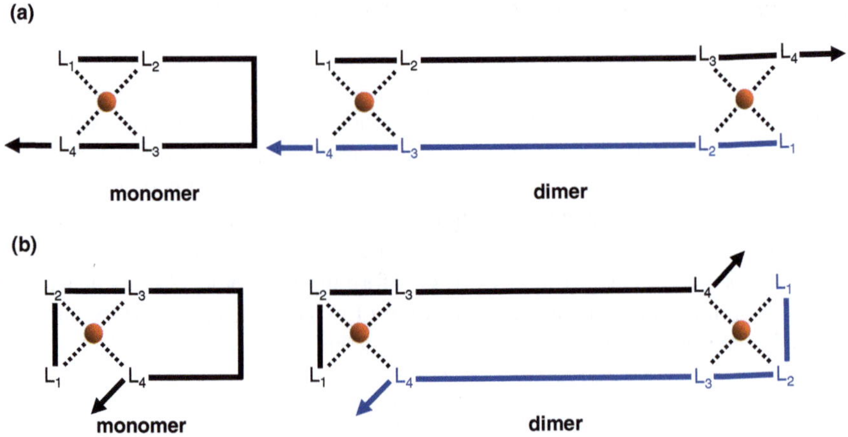

Fig. 1 Schematic depicting the local and quaternary structure of a peptide coordinating a metal ion (*brown sphere*) in a 4-coordinate planar fashion. The individual ligands (L_1, …, L_4) involved in the first coordination sphere (either backbone or side chain) are shown explicitly, while the remaining peptide backbone is drawn as a solid line. Without introduction of isotopic labels, the first coordination sphere of monomer and dimer cannot be distinguished. (**a**) A symmetric bridging interaction between antiparallel peptides wherein the first coordination sphere contains ligands L_1, L_2 from one subunit and L_2, L_3 from another: $\{L_1, L_2 \mid L_3, L_4\}$. (**b**) A symmetric bridging interaction with the first coordination sphere containing ligands L_1, L_2, L_3 from one subunit and L_4 from another: $\{L_1, L_2, L_3 \mid L_4\}$. The direction of the peptide chain (from N- to C-terminus) is indicated by the *arrow*. Monomer and dimer structures are drawn, although higher order oligomers M_nP_n share the same bridging interaction as the dimers, albeit with different quaternary structure

metals with more than one naturally-abundant isotope, such as copper (69% ^{63}Cu, 31% ^{65}Cu), the different nuclear magnetic moments [10] result in an offset of the shf pattern for each isotope and thus 'destructive interference'; therefore, resolution of shf lines is improved by use of mono-isotopic metal ion. For frozen solution EPR spectra, inclusion of glycerol in the range 0–20% v/v can improve resolution of shf structure by preventing separation of the solution into solute and purified ice phases, although glycerol is also a modest osmolyte [19] and may therefore affect protein folding and protein-protein interactions. For metal-bridged low-molecular-weight systems, it may also be possible to examine the shf structure using solution phase EPR spectra obtained at room temperature.

Numerical Simulation

The spin Hamiltonian describing the energy of an individual Cu^{2+} coordination site is given by:

$$H = \beta \mathbf{B} \cdot \mathbf{g} \cdot \mathbf{S} + \mathbf{S} \cdot \mathbf{A} \cdot \mathbf{I} - g_n \beta_n \mathbf{B} \cdot \mathbf{I} + \sum_k \left(\mathbf{S} \cdot \mathbf{A}(L_k) \cdot \mathbf{I}_k - g_{n,k} \beta_n \mathbf{B} \cdot \mathbf{I}_k \right) \quad (2)$$

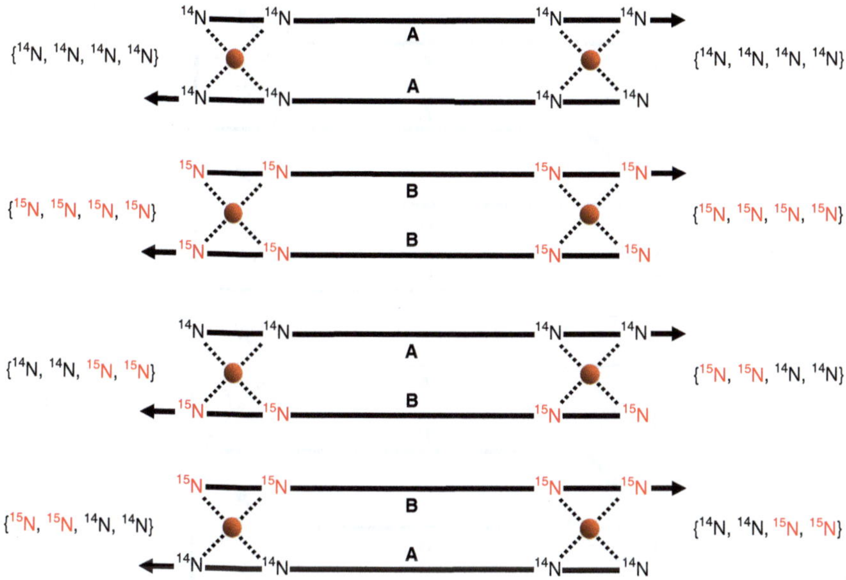

Fig. 2 Illustration of all equally possible structures formed by the antiparallel dimer {L_1, L_2 I L_3, L_4}={N, N I N, N} for an inhomogeneous solution comprised of M/A/B 1:½:½, where peptide A is unlabelled and peptide B has L_1=^{15}N, L_2=^{15}N, L_3=^{15}N, L_4=^{15}N (or uniform ^{15}N-labelling)

where **S** and **I** are the electron and nuclear vector spin operators, **g** is the 3 × 3 electron Zeeman matrix, **A** is the hyperfine coupling matrix associated with the Cu nucleus, β is the Bohr magneton, β_n is the nuclear magneton and **B** is the applied magnetic field. The summation incorporates the shf and nuclear Zeeman interactions of the kth ligand nucleus of spin I_k and nuclear g factor $g_{n,k}$. Rhombic symmetry was permitted for the electron Zeeman and ^{65}Cu hyperfine interaction, where required. The hyperfine matrices **A** are proportional to the g factor of the relevant nucleus. Thus, **A**(^{65}Cu) = **A**(^{63}Cu) × g_n(^{65}Cu)/g_n(^{63}Cu) = **A**(^{63}Cu) × 1.07 and **A**(^{15}N) = **A**(^{14}N) × |g_n(^{15}N)/g_n(^{14}N)| = **A**(^{14}N) × 1.40 [10]. Oxygen-16 has no nuclear magnetic moment (I = 0), although ^{17}O (I=5/2) can be used. Quadrupole terms for the nuclei are neglected in Eq. (2) but may be required in some instances.

To illustrate the approach Cu^{2+}-bridged dimers are assumed and a hypothetical set of spin Hamiltonian parameters (Table 3) has been used and the low frequency (4.0 GHz) EPR spectrum simulated with Easyspin v.5.0.2 [20] utilizing the 'pepper' function to solve the spin Hamiltonian (1) *via* a second order perturbation expression (using the "perturb2" method). The **g** and **A**(Cu) matrices were assumed to be axially symmetric with collinear principal directions. Guassian distributions of the principal g_i and A_i values (i = x, y, z) were included using a linewidth model incorporating **g** and **A**(Cu) strain [18], with each of the principal g_i and A_i values having a positive correlation. For simplicity, the metal-ligand shf coupling of the kth ligand

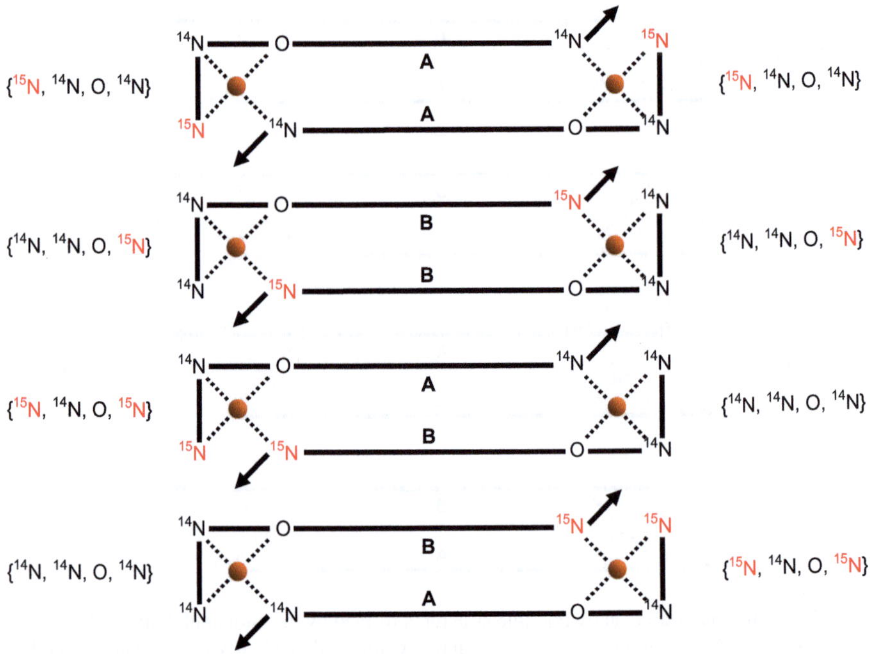

Fig. 3 Illustration of all equally possible structures formed by the antiparallel dimer $\{L_1, L_2, L_3 \mid L_4\}=\{N, N, O \mid N\}$ for an inhomogeneous solution comprised of M/A/B 1:½:½, where peptide A has site-selective $L_1={}^{15}N$ and peptide B has site-selective $L_4={}^{15}N$

Table 1 Relative speciation of the four first coordination spheres that can be formed by the antiparallel dimer $\{L_1, L_2 \mid L_3, L_4\}$ shown in Fig. 1a, for a limited selection of combinations of labelled ligand nuclei. An inhomogeneous solution of M/A/B 1:½:½ is assumed, where A and B have the same peptide but with different combinations of isotopic labels (*) on coordinating residues

Peptide mixture	First coordination sphere	Relative Speciation	
		Monomer	Oligomer
Peptide A: $*L_1, L_2, L_3, L_4$ Peptide B: $L_1, *L_2, L_3, L_4$	$\{*L_1, L_2, L_3, L_4\}$	½	½
	$\{L_1, *L_2, L_3, L_4\}$	½	½
Peptide A: $*L_1, L_2, L_3, L_4$ Peptide B: $L_1, L_2, *L_3, L_4$	$\{*L_1, L_2, L_3, L_4\}$	½	¼
	$\{L_1, L_2, *L_3, L_4\}$	½	¼
	$\{L_1, L_2, L_3, L_4\}$	0	¼
	$\{*L_1, L_2, *L_3, L_4\}$	0	¼
Peptide A: L_1, L_2, L_3, L_4 Peptide B: $*L_1, *L_2, *L_3, *L_4$	$\{L_1, L_2, L_3, L_4\}$	½	¼
	$\{*L_1, *L_2, *L_3, *L_4\}$	½	¼
	$\{L_1, L_2, *L_3, *L_4\}$	0	¼
	$\{*L_1, *L_2, L_3, L_4\}$	0	¼

Table 2 Relative speciation of the four first coordination spheres that can be formed by the antiparallel dimer $\{L_1, L_2, L_3 \mid L_4\}$ shown in Fig. 1b, for a limited selection of combinations of labelled ligand nuclei. An inhomogeneous solution of M/A/B 1:½:½ is assumed, where A and B have the same peptide sequence but with different combinations of isotopic labels (*) on coordinating residues

Peptide mixture	First coordination sphere	Relative Speciation	
		Monomer	Oligomer
Peptide A: $*L_1, L_2, L_3, L_4$ Peptide B: $L_1, *L_2, L_3, L_4$	$\{*L_1, L_2, L_3, L_4\}$	½	½
	$\{L_1, *L_2, L_3, L_4\}$	½	½
Peptide A: $*L_1, L_2, L_3, L_4$ Peptide B: $L_1, L_2, L_2, *L_4$	$\{*L_1, L_2, L_3, L_4\}$	½	¼
	$\{L_1, L_2, L_3, *L_4\}$	½	¼
	$\{*L_1, L_2, L_4, *L_4\}$	0	¼
	$\{L_1, L_2, L_3, L_4\}$	0	¼
Peptide A: L_1, L_2, L_3, L_4 Peptide B: $*L_1, *L_2, *L_3, *L_4$	$\{L_1, L_2, L_3, L_4\}$	½	¼
	$\{*L_1, *L_2, *L_3, *L_4\}$	½	¼
	$\{L_1, L_2, L_3, *L_4\}$	0	¼
	$\{*L_1, *L_2, *L_3, L_4\}$	0	¼

Table 3 Spin Hamiltonian parameters used to demonstrate the isotopic dilution method. Simulations for specific examples given in Figs. 2 and 3 are shown in Figs. 4 and 5, respectively. Isotropic ligand hyperfine couplings $a(L_k)$ for each ligand k are given for ^{14}N, but were scaled by $|g_n(^{15}N)/g_n(^{14}N)| = 1.40$ for each labelling scheme as appropriate

Example	Electron Zeeman		Hyperfine (^{65}Cu) (MHz)		Ligand hyperfine (^{14}N) (MHz)			
	$g_{\parallel}$	$g_{\perp}$	$A_{\parallel}$	$A_{\perp}$	$a(L_1)$	$a(L_2)$	$a(L_3)$	$a(L_4)$
{N, N I N, N} (Fig. 4)	2.25	2.06	600	60	33	39	39	36
{N, N, O I N} (Fig. 5)	2.25	2.06	600	60	33	39	–	36

Residual linewidth (H strain): $\sigma_{R_x} = \sigma_{R_y} = \sigma_{R_z} = 25\,\text{MHz}$ (FWHM)

g strain: $\sigma_{g_x} = \sigma_{g_y} = 0.02, \sigma_{g_z} = 0.04$

A strain: $\sigma_{A_x} = \sigma_{A_y} = \sigma_{A_z} = 5\,\text{MHz}$ (FWHM)

nucleus was assumed isotropic. ie. $A_x(L_k) = A_y(L_k) = A_z(L_k) = a(L_k)$. The magnitudes of $a(L_1)$, $a(L_2)/a(L_3)$ and $a(L_4)$ in Table 3 are representative of hypothetical couplings to an N-terminal amino, backbone amide(s) and an imino nitrogen of His side chain, respectively. The effect of field modulation was emulated by setting the 'ModAmp' parameter in Easyspin to 5 gauss. All spectra were normalized by their integrated absorption over the entire spectral range.

Examples

For the quaternary M_2P_2 structures shown in Fig. 1, a selected number of possible labelling schemes, together with the first coordination spheres that can be formed by a mixture Cu/A/B 1:½:½, are listed in Table 1 and Table 2. To illustrate the effective application of isotopic dilution, two specific examples relevant to Cu^{2+}-bridged peptide oligomers are provided, both of which use a common set of spin Hamiltonian parameters listed in Table 3. In the following, the notation $\{L_1, L_2, L_3, L_4\}$ refers to a first coordination sphere with equatorial ligands L_1–L_4, while a vertical bar (|) separating two ligands is used to indicate this coordination is intermolecular.

The structure in Fig. 1a forms two structurally identical Cu_2P_2 binding sites with $\{L_1, L_2 \mid L_3, L_4\}$ coordination. For the first site, one peptide subunit supplies ligands L_1 and L_2 and the other supplies ligands L_3 and L_4, while the contribution of each peptide is reversed for the other site. Let us assume a 4N coordination sphere and a mixture Cu/A/B 1:½:½ containing unlabelled peptide A and uniformly ^{15}N-labelled peptide B. In this instance, contributions from $\{^{15}N,^{15}N,^{15}N,^{15}N\}$ and $\{^{14}N,^{14}N,^{14}N,^{14}N\}$ coordination modes are expected for monomeric Cu^{2+} binding, with a spectrum dominated by the intense modulation depth of the $4\times^{15}N$ shf pattern of $\{^{15}N,^{15}N,^{15}N,^{15}N\}$. When Cu^{2+}-bridging occurs, the additional signals from $\{^{15}N,^{15}N,^{14}N,^{14}N\}$ and $\{^{14}N,^{14}N,^{15}N,^{15}N\}$ coordination modes (Fig. 2) leads to a clear difference in the intensity and number of shf lines (Fig. 4).

In the second example, two identical Cu^{2+} binding sites with $\{L_1, L_2, L_3 \mid L_4\}$ coordination are formed from antiparallel peptide chains, where one peptide supplies three ligands and the other supplies the fourth (Fig. 1b). Assume now a 3N1O coordination and a solution Cu/A/B 1:½:½ containing peptides with site-selective L_1=^{15}N (peptide A) and site-selective L_4=^{15}N (peptide B). Monomeric Cu^{2+} binding produces only $\{^{15}N,^{14}N, O,^{14}N\}$ and $\{^{14}N,^{14}N, O,^{15}N\}$, whose spectra yield comparable $1\times^{15}N/2\times^{14}N$ shf patterns. Metal bridging introduces additional coordination from $\{^{14}N,^{14}N,O,^{14}N\}$ and $\{^{15}N,^{14}N, O,^{15}N\}$ (Fig. 3), whose shf patterns have intermediate line spacing that leads to destructive interference and a clear loss of modulation depth (Fig. 5).

Depending on the type of bridging, the most effective scheme may involve either site-selective or uniform labelling. Using a mixture of unlabelled and uniformly labelled protein is suitable for large proteins where the introduction of site-selective isotopic labels by solid phase peptide synthesis would be impractical or impossible, although site-selective labelling may result in a clearer delineation of monomeric and oligomeric shf patterns in some instances. When using site-selective labelling, one must ensure the chosen labelling scheme is capable of distinguishing oligomeric from monomeric metal binding. For example, if peptide A has a site-selective label on L_1 (denoted $^*L_1$) and peptide B has a site-selective label on L_2 (denoted $^*L_2$), it can be seen for the given examples that only $\{^*L_1, L_2, L_3, L_4\}$ and $\{L_1, ^*L_2, L_3, L_4\}$ coordination spheres will result, making the CW-EPR spectrum indistinguishable from that expected for monomeric Cu^{2+} binding (Table 1, Table 2). For the same reasons, when the first coordination sphere contains only a single nitrogen ligand, or when nitrogen ligands appear only on one side of the bridging metal ion (e.g. an antiparallel dimer with {N, N | O, O} coordination), significantly more expensive site-selective ^{17}O-labelling would be necessary to distinguish the shf patterns of monomers and metal-bridged oligomers.

Experimental confirmation of intermolecular metal bridging relies on the identification of a clear difference between the EPR spectra of [M/A 1:1 + M/B 1:1] and

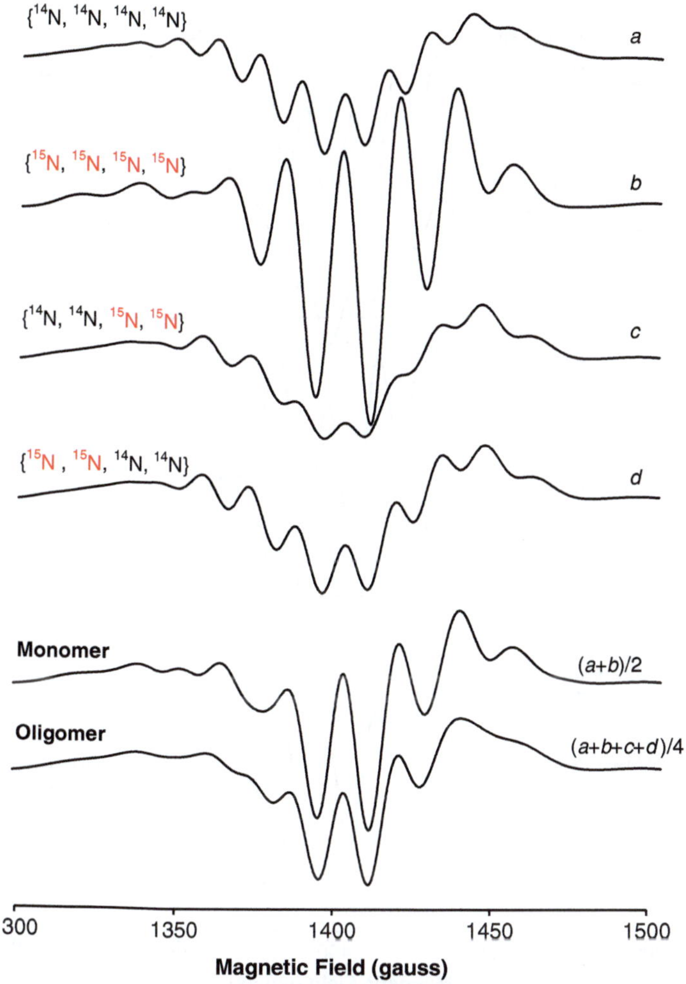

Fig. 4 Spectral simulations of the second derivative S-band EPR spectra of the model system depicted in Fig. 1a, with $\{L_1, L_2 \mid L_3, L_4\} = \{N, N \mid N, N\}$. Peptide A is unlabelled and peptide B is uniformly ^{15}N labelled. In the absence of metal-bridging of peptide monomers, a solution of Cu/A/B 1:½:½ contains a simple average of the spectra of (a) $\{^{14}N,^{14}N,^{14}N,^{14}N\}$ (Cu/A 1:1) and (b) $\{^{15}N,^{15}N,^{15}N,^{15}N\}$ (Cu/B 1:1). When metal bridging of peptide monomers occurs, additional first coordination spheres (c) $\{^{15}N,^{15}N,^{14}N,^{14}N\}$ and (d) $\{^{14}N,^{14}N,^{15}N,^{15}N\}$ are possible (Table 1, Fig. 2)

M/A/B 2:1:1. The presence of multiple species in solution (e.g. mixture of monomers and bridged oligomers and/or overlapping pH-dependent coordination modes) can obscure the above difference and hence make the presence and structure of any bridged oligomer unclear. Even if the above difference is large enough to clearly identify a metal bridging interaction, knowledge of the spin Hamiltonian parameters (local structure) may help to further ascertain the quaternary structure associated with the bridging (Fig. 1). Good estimates of these parameters should be obtainable

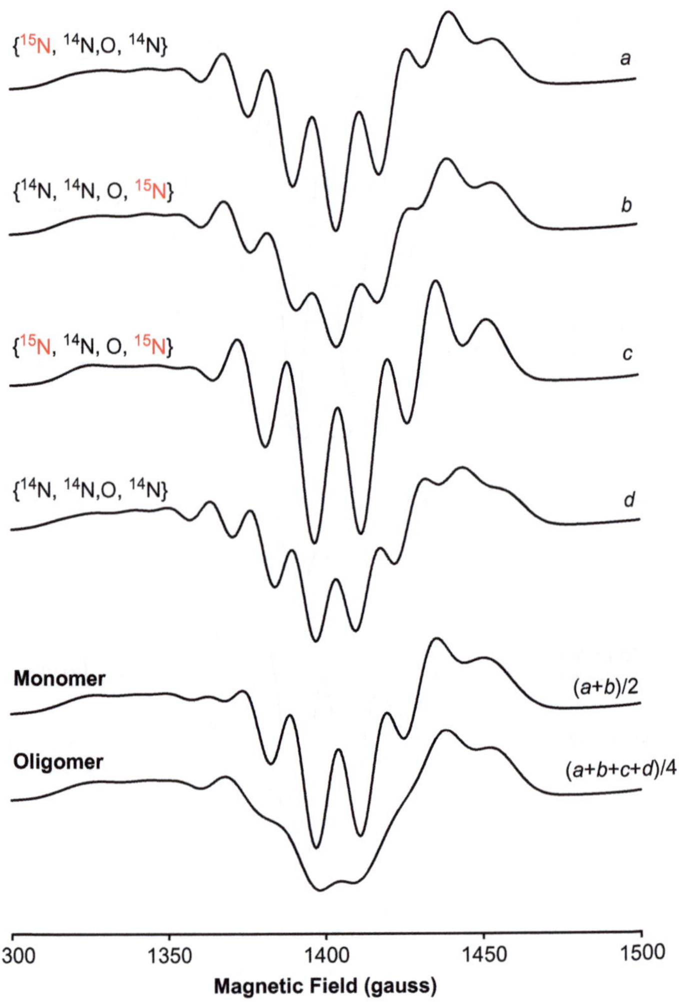

Fig. 5 Spectral simulations of the second derivative S-band EPR spectra of the model system depicted in Fig. 1b, with {L_1, L_2, L_3 | L_4} = {N, N, O | N}. Peptide A has site-selective L_1=^{15}N and peptide B has site-selective L_4=^{15}N. In the absence of metal-bridging of peptide monomers, a solution of Cu/A/B 1:½:½ contains a simple average of the spectra of (a) {^{15}N, ^{14}N, O, ^{14}N} (Cu/A 1:1) and (b) {^{14}N, ^{14}N, O, ^{15}N} (Cu/B 1:1). When metal bridging of peptide monomers occurs, additional first coordination spheres (c) {^{14}N, ^{14}N, O, ^{14}N} and (d) {^{15}N, ^{14}N, O, ^{15}N} are formed (Table 2, Fig. 3)

from numerical simulation of the experimental spectra of unlabelled Cu/P 1:1, together with those of labelled M/A 1:1 and M/B 1:1. These parameters should then provide reliable values to simulate the shf pattern of coordination modes for which there may be no experimental data available (e.g. the {^{15}N, ^{14}N, O, ^{15}N} mode in Fig. 5) and thus to predict the shf pattern associated with different kinds of quaternary structures. If there is any doubt as to the validity of the spin Hamiltonian parameters, one may synthesise additional peptides containing two site-selective isotopic labels (e.g. one with $L_1 = {}^{15}N$, $L_2 = {}^{14}N$, $L_3 = O$, $L_4 = {}^{15}N$ for the example given in Fig. 5) to enable the acquisition of a complete set of experimental EPR spectra representing all four possible coordination spheres in Tables 1 and 2. Linear combinations of these experimental spectra can then be compared in an analogous manner to Figs. 4 and 5 in order to identify the quaternary structure without recourse to numerical simulation.

Finally, some limitations should be mentioned. Firstly, the method does not distinguish between passive (metal bridging takes place after oligomerisation) and active (metal bridging drives oligomerisation) metal binding. Secondly, although examples of two dimeric structures are provided, the relative speciation in Tables 1 and 2 is also predicted for higher order closed-chain oligomers M_nP_n of any order, as well as for open chain bridged oligomers M_mP_n ($m = n-1$) [7]. In the latter case, open-chain metal bridging might be identified by a metal binding saturation point at $[M]/[P] = (n-1)/n$ as compared with $[M]/[P] = n$ for a closed-chain, although the distinction becomes more difficult to ascertain with confidence for increasing chain size n. Finally, the methodology can, in principle, also be applied to mononuclear mixed-ligand (ternary) metal complexes MA_jB_k of coordination number $n = j + k$. Each of the possible first coordination spheres will no longer be geometrically equivalent, however, and therefore their speciation will deviate from 0.25 by an amount determined by their relative stability [1, 2].

References

1. Sigel H (1975) Angew Chem Int Ed 14:394–402
2. Sharma VS, Schubert J (1969) J Chem Educ 46:506–507
3. Drew SC (2015) Appl Magn Reson 46:1041–1052
4. Pedersen JT, Teilum K, Heegaard NHH, Østergaard J, Adolph H-W, Hemmingsen L (2011) Angew Chem Int Ed 50:2532–2535
5. Drew SC (2015) Chem Eur J 21:7111–7118
6. Hayes RG (1969) Ligand hyperfine couplings and the structure of transition metal complexes. In: Yen TF (ed) Electron spin resonance of metal complexes. Plenum Press, New York, pp 23–32
7. Drew SC (2016) J Inorg Biochem 158:5–10
8. Petkova AT, Yau W-M, Tycko R (2006) Biochemistry 45:498–512
9. Peisach J, Blumberg WE (1974) Arch Biochem Biophys 165:691–698
10. Stone NJ (2005) At Data Nucl Data Tables 90:75–176
11. Burns CS, Aronoff-Spencer E, Dunham CM, Lario P, Avdievich NI, Antholine WE, Olmstead MM, Vrielink A, Gerfen GJ, Peisach J, Scott WG, Millhauser GL (2002) Biochemistry 41:3991-–4001
12. Burns CS, Aronoff-Spencer E, Legname G, Prusiner SB, Antholine WE, Gerfen GJ, Peisach J, Millhauser GL (2003) Biochemistry 42:6794–6803
13. Drew SC, Noble CJ, Masters CL, Hanson GR, Barnham KJ (2009) J Am Chem Soc 131:1195–1207
14. Drew SC, Barnham KJ (2011) Acc Chem Res 44:1146–1155

15. Schweiger A, Jeschke G (2001) Principles of pulse electron paramagnetic resonance. Oxford University Press, Oxford
16. Palmer G (2000) Electron paramagnetic resonance of metalloproteins. In: Que L (ed) Physical methods in bioinorganic chemistry: spectroscopy and magnetism. University Science Books, California, p 213
17. Pilbrow JR (1990) Transition ion electron paramagnetic resonance. Clarendon Press, Oxford
18. Hyde JS, Froncisz W (1982) Annu Rev Biophys Bioeng 11:391–417
19. Street TO, Bolen DW, Rose GD (2006) Proc Nat Acad Sci USA 103:13997–14002
20. Stoll S, Schweiger A (2006) J Magn Reson 178:42–55

Structures, Electronics and Reactivity of Copper(II) Complexes of the Cyclic *Pseudo*-Peptides of the Ascidians

Peter Comba and Annika Eisenschmidt

Abstract Cyclic pseudo-peptides, derived from marine metabolites found in the genus *Lissoclinum bistratum* and *Lissoclinum patella* and produced by their symbiont *Prochloron* have attracted scientific interest in the last two decades. Their structural properties and solution dynamics were analyzed in detail, elaborate synthetic procedures for the natural products and synthetic derivatives were developed, the biosynthetic pathways were studied and it now is possible to produce them biosynthetically. A major focus in the last decade was on their Cu^{II} – more recently also on the Zn^{II} – coordination chemistry, as a number of studies have indicated that dinuclear Cu^{II} and Zn^{II} complexes of cyclic peptides may be involved in the ascidians' metabolism. Solution equilibria of various derivatives of the synthetic ligands in presence of Cu^{II} were studied thoroughly and the solution structures were determined by EPR spectroscopy and spectra simulation in combination with molecular mechanics and DFT calculations. Recent in vitro studies indicate that the dicopper(II) complexes are phosphatase, glycosidase, lactamase and very efficient carbonic anhydrase model systems. First in vivo studies with a patellamide derivative containing an appended fluorescence tag suggest that Cu^{II} is coordinated to the patellamides in the *Prochloron* cells.

Keywords *Prochloron* • Patellamide • Carboanhydrase • Dipole–dipole coupling • Solution structure • Cyclic peptide

Part of this chapter is reproduced from the PhD thesis of A.E.

P. Comba (✉) • A. Eisenschmidt
Anorganisch-Chemisches Institut and Interdisciplinary Center for Scientific Computing (IWR), Ruprecht-Karls Universität Heidelberg, Im Neuenheimer Feld 270, 69120 Heidelberg, Germany
e-mail: peter.comba@aci.uni-heidelberg.de

G. Hanson, L. Berliner (eds.), *Future Directions in Metalloprotein and Metalloenzyme Research*, Biological Magnetic Resonance 33,
DOI 10.1007/978-3-319-59100-1_2

Background

Prochloron: *A Special Cyanobacterium*

The International Union of Geological Sciences (IUGS) is the body responsible for the official geological periodization. To IUGS, the atmospheric scientists PAUL CRUTZEN and EUGENE STOERMER put forward the suggestion to call the era from the 1960s onwards 'Anthropocene' [1, 2]. If the commission decides in favor, it would mean that humans consider themselves as predominantly responsible for the sedimentation observed on earth for the last approximately 50 years.

Without the emergence of oxytrophic organisms, namely cyanobacteria, more than 2.5 billion years ago, there would not be an Anthropocene, simply because of the lack of air to breathe—there would simply not be sufficient dioxygen. The first organisms to produce dioxygen, as a very toxic waste product from the metabolic conversion of CO_2, were cyanobacteria. This development was driven by the need of reduced carbon species for the growth of organisms, the required amount of oxidation equivalents could not be buffered any longer by auxiliary substrates like reduced iron or sulfur species and dioxygen was directly released to the atmosphere [3]. Due to their photosynthetic dioxygen production, cyanobacteria are to date the only reported examples of oxyphototrophs among prokaryotes and are mainly responsible for the present composition of the biosphere and oxic atmosphere [4]. It also appears that the higher plants acquired their efficient machineries by the incorporation of prokaryotes as symbionts. This is commonly referred to as theory of symbiogenesis [5], and was first proposed in 1910 but did not gain much attention for more than half a century [6–8], when the unicellular cyanobacteria as models for the prokaryotic ancestor of red algal chloroplasts were found, and finally, *Prochloron* the potential model ancestor for the green algal chloroplast was discovered [8–10]. This photosynthetic symbiont of ascidians exhibits not only chlorophyll *a*, but also chlorophyll *b*, allowing a more efficient collection of sun light, as in eukaryotes, i.e. higher plants and algae. However, unlike other oxygenic photosynthetic prokaryotes *Prochloron* does not have blue or red protein pigments (phycobilins).

Prochloron is an extremely large prokaryote, with a diameter of 7–25 μm (see Fig. 1). It is a spherical cell with thylakoid membranes stacked to the periphery [4].

Fig. 1 *Left*: Microscopy picture of *Prochloron*, showing the stacked thylakoid membranes inside (taken with a confocal microscope λ_{exc} = 665 nm); *middle*: *Lissoclinum patella* colony ~200 cm^2 on a coral; *right*: ~10 cm^2 on a stone

It is always found as an obligate symbiont of subtropical and tropical ascidians, like *Diplosoma virens, Tridemnum cyclops, Didemnum molle* and the genus *Lissoclinum–Lissoclinum punctatum, Lissoclinum bistratum* and *Lissoclinum patella* amongst them. *Prochloron* is the only reported obligate photosymbiont in the phylum chordata [11]. This is particularly interesting, because a *Prochloron* draft genome was published [12] that did not show any lack of relevant metabolic genes that could explain why it can neither sustain photosynthesis nor reproduce *ex hospite* (outside the host). Currently research is being undertaken to unravel the secrets of *Prochloron*'s precise microenvironment and consequently understand the requirements to transfer it to laboratory culture [4].

For carbon assimilation, *Prochloron*, like other blue-green algae, uses carboxysomes [13], in which carbonic anhydrase and Rubisco are abundant to fix CO_2 to phosphoglyceric acid. With an oxygen evolving complex and electron transport observed via photosystem I and II, *Prochloron*'s photosynthetic apparatus appears related to that of other cyanobacteria. The putative fixation of N_2 by *Prochloron* is subject of ongoing discussion [12, 14]. More recent studies focused on the *Prochloron* genome, the microbiome as well as on the micro-environment and added significantly to the understanding of the ecology of *Prochloron* [12, 15, 16]. Importantly, it was demonstrated that *Prochloron* is subject to rapid and dynamic chemical changes in its direct environment: [4] upon irradiation, the pH as well as the O_2 saturation levels are altered drastically within 15–30 min—changing the interior zone of the ascidians from an anoxic and neutral/slightly acidic environment (pH ~7.0) to a dioxygen super saturation level and strongly alkaline pH values (10.5) compared to the surrounding sea water (pH 8.2) [4]. Surprisingly, no photo inhibition was observed, even upon irradiance levels comparable to direct sunlight, and the photosynthesis recovered quickly after periods of extremely low dioxygen saturation levels [4]. Since CO_2 is absent at these alkaline conditions and photosynthesis consequently is solely dependent on HCO_3^-, this points to an HCO_3^- dependent carbon transport mechanism [17].

To what extent the biochemical and photosynthetic activity of *Prochloron* is influenced by the host is uncertain to date and subject of current research [13, 18]. Implications on possible communication mechanisms across the cell membrane arose from secondary metabolites found inside *Prochloron* as well as in close proximity to it, in the ascidians. These secondary metabolites are small cyclic peptides, that are ribosomally expressed and are commonly referred to as cyanobactins [12, 19, 20].

Ribosomal Expression of Cyclic Peptides

Prochloron exhibits a rich variety in metabolites that are apparently expressed depending on which host it is living in symbiosis with. The cyclic peptides were first discovered during the 1980s [21]. Medicinal applications of the natural cyclic peptides and a series of synthetic derivatives have been studied thoroughly and it was shown that these peptides exhibit cytotoxic [21], antibacterial, antineoplastic and antiviral activities [22, 23]. Some of the metabolites found in host organisms of *Prochloron* are summarized in Chart 1.

The cyanobactins found in *L. patella*, ascidiacylclamide, patellamides A-G and ulicyclamide, are all expressed via the ribosomal production and post-translational modification (the RIPP mechanism) [23–25]. As this mechanism is usually only observed for peptides of metabolic importance, it gives rise to the hypothesis that the cyclic peptides might act as key metabolites [24, 26, 27].

Common chemical features of the cyanobactins are the heterocycles, i.e. thiazole, oxazole, as well as oxazoline, which are connected via peptide bonds. The amino acid residues are hydrophobic, and often originate from valine or *iso*-leucine. *N*-heterocycles result from the condensation of threonine, cysteine and serine side chains with carbonyl groups in the peptide sequence [35]. Condensation of six or eight amino acids results in hexa- or octapeptides, and since the peptide functionalities are hidden in the heterocycle, the macrocycles are called *pseudo*-peptides (see Chart 1 for highlighted constituents and the corresponding amino acids and Table 1 for the side chains of patellamide A–G).

The *pseudo*-octapeptides isolated from *L. patella* show a strong preorganization for the coordination of metal ions with their 21-azacrown-7 and 24-azacrown-8 structures. Interestingly, none of the natural peptides exhibit imidazole heterocycles

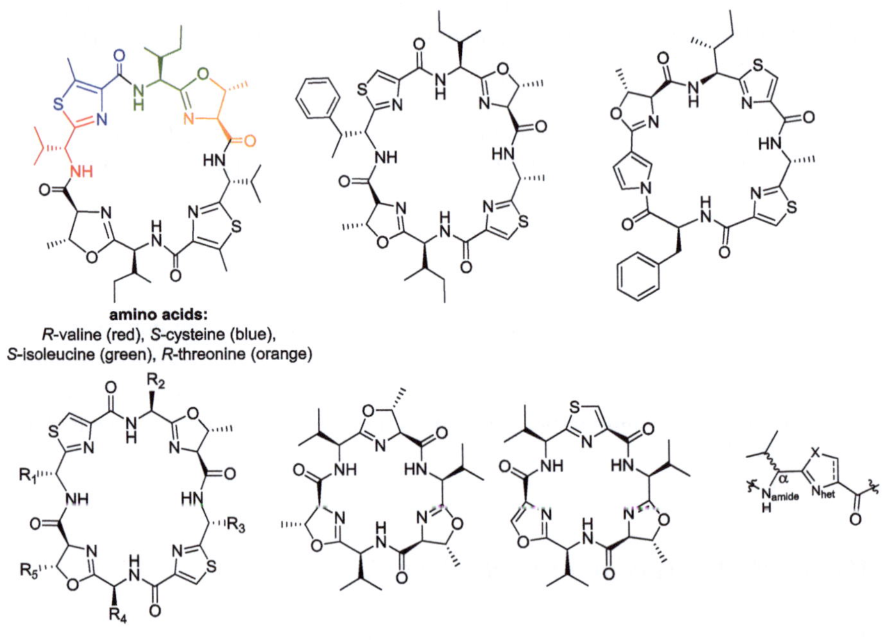

Chart 1 Ascidiacyclamide [28], Patellamide A [29], Ulicyclamide [30], Patellamide skeleton (see also Table 1), Westiellamide [31], Bistratamide D [32], schematic representation of one cyanobactin building block (*left* to *right*, *top* to *bottom*). The color code in ascidiacyclamide shows the amino acid building blocks of one half of the molecule [33–35]

Table 1 Amino acids contributing to the side chains of Patellamide A–G [29]

Patellamide	R_1	R_2	R_3	R_4	R_5
A	R-Val	S-Ile	R-Val	*S*-Ile	H
B	R-Phe	S-Ile	R-Ala	S-Leu	CH_3
C	R-Phe	S-Ile	R-Ala	S-Val	CH_3
D	R-Phe	S-Ile	R-Ala	S-Ile	CH_3
E	R-Phe	S-Ile	R-Val	S-Val	CH_3
F	R-Val	S-Val	R-Phe	S-Val	H
G	R-Ala	S-Leu	R-Phe	S-Ile	CH_3

Scheme 1 Preferred conformations of cyclic *pseudo*-peptides [38, 39]

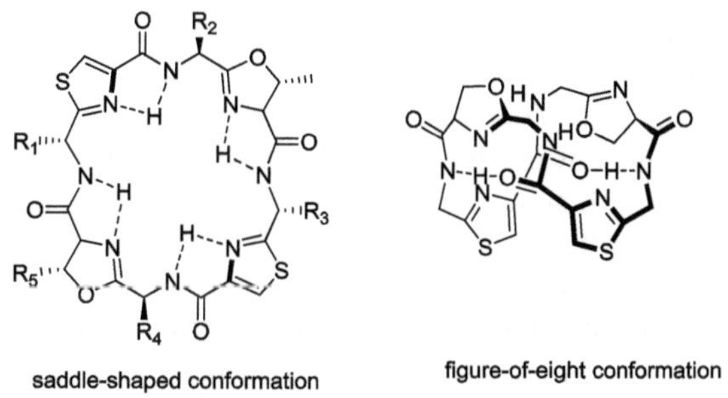

that are commonly conserved in metal ion binding sites of metabolic importance [3]. Another common feature in the cyclic octapeptides is the alternate *R,S,R,S* configuration of the side chains, leading to a limited flexibility of the macrocycles.

The rigidity of the systems was investigated by density functional theory (DFT), showing that it is primarily governed by the incorporated heterocycle [36, 37]. Analyses of the octapeptides by means of DFT calculations, in addition to X-ray crystallography and ^{1}H-NMR spectroscopy imply two different conformations: the saddle-shaped structure and the twisted 'figure-of-eight' conformation (see Scheme 1) [38].

Less is known about the cyclic hexapeptides, like the bistratamides (see Chart 1), which could be extracted from *Lissoclinum bistratum* [40]. Although the exact expression mechanism was not elucidated, it is believed that *Prochloron*, the symbiont of *L. bistratum*, expresses the cyclic peptides as observed in the *L. patella* symbiosis [40]. Very closely related hexapeptides were found from the extraction of the terrestrial blue-green alga *Westiellopsis prolifica* ("The blue-green alga was collected from a mud sample" [31]), called westiellamides (see Chart 1) [31]. The structural difference between westiellamides and bistratamides is mainly the incorporation of a thiazole instead of an oxazoline.

Given their high bioactivity, the cyclic peptides might play an important role in anti-predator mechanisms. However, this is equally disputable, because of the sheer quantity of cyanobactins found in *Lissoclinum patella* and *Lissoclinum bistratum*—up to several percent of the dry weight [24, 26]. Consequently, the current focus of research is on the putative metabolic role of the cyclic peptides. The significantly increased metal ion concentration, in particular of Cu^{II} and Zn^{II} to about 10-fold[4] in comparison to the surrounding sea water implied a metal-coordination related metabolic role [41]. Different functions have been discussed, the involvement in metal ion transport and storage as well as their role as potential prosthetic groups amongst them. Therefore, the coordination chemistry of natural cyclic hexa- and octapeptides was studied and is presented in the next paragraph [38, 41].

Metal Complexes of Cyclic Hexa- and Octapeptides

The metal ion concentrations (Cu^{II} and Zn^{II}) found in ascidians are comparably high, which is surprising if compared to the cytotoxic effects known at these concentrations for other microorganisms [41]. Given the constitution of the patellamides with four heterocyclic-nitrogen atoms and four amide-nitrogen atoms, pointing towards the inside of the cyclic peptide, potentially acting as donor atoms, the coordination chemistry of these macrocycles has been studied extensively in vitro [38].

ITC measurements, as well as data from CD, UV-vis and NMR spectroscopy indicate a metal ion binding of both Cu^{II} as well as Zn^{II} with stability constants in the range of $\log K(Zn^{II}) \approx 2$–$4$, $\log K(Cu^{II}) \approx 4$–$5$ (data given here were determined by CD-spectroscopy in methanolic solutions) [38]. The X-ray structure of the dinuclear Cu^{II} complex of ascidiacylamide with carbonate as a bridging coligand was the first natural metal patellamide complex that could be investigated in the solid state (and is shown in Fig. 2) [41]. In this complex, two metal ions are binding to the ligand, the Cu^{II} is coordinated by two heterocyclic nitrogen donors as well as a deprotonated amide nitrogen, providing a N_{het}-N_{amide}-N_{het} binding motif. This is accompanied by acidification of the solution, since protons are released from the ligand upon complexation. Consequently, two equivalents of base have to be added

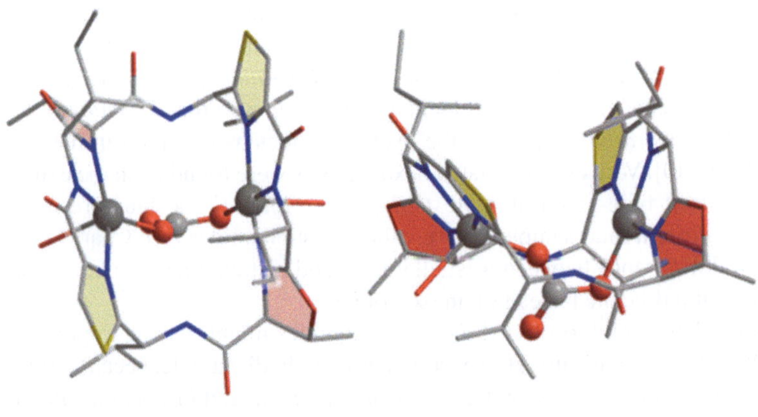

Fig. 2 Top and side view of the carbonato-bridged dinuclear Cu^{II} complex of ascidiacyclamide determined by X-ray crystallography (*grey*: C, *yellow*: S, *blue*: N, *red*: O) [41]. Reprinted with permission from [34] Copyright 2012 Wiley-VCH

in order to achieve the formation of a dinuclear complex [42]. As the pH in sea water is approximately 8.2, which is slightly alkaline, it is not unlikely that the dinuclear Cu^{II} patellamide complexes are associated with the metabolic role of the cyclic octapeptides.

An investigation of the metal-binding capability of patellamide D by means of EPR, CD and UV-vis titrations indicated the following complexation scenario: [41]

$$[Cu(MeOH)]^{2+} + \mathbf{H_4patD} + NEt_3 \longrightarrow [Cu(\mathbf{H_3patD})]^{+} + NEt_3H^{+}$$

$$[Cu(\mathbf{H_3patD})]^{+} + (Cu^{2+})_{solv.} + NEt_3 \longrightarrow [Cu_2(\mathbf{H_2patD})]^{2+} + NEt_3H^{+}$$

$$[Cu_2(\mathbf{H_2patD})]^{2+} + NEt_3 + H_2O \longrightarrow [Cu_2(\mathbf{H_2patD})(OH)]^{+} + NEt_3H^{+}$$

$$[Cu_2(\mathbf{H_2patD})(OH)]^{+} + CO_2 + H_2O \longrightarrow [Cu_2(\mathbf{H_2patD})(CO_3)] + H_3O^{+}$$

Interestingly, the dinuclear Cu^{II} patellamide complexes with OH^- as the coligand are observed to form carbonato-bridged complexes upon exposure to air, as shown for the corresponding ascidiacyclamide complex (see Fig. 2) [34].

Model Peptides

Obtaining patellamides from natural sources is not only invasive, but also time-consuming and delivers comparably small amounts of the peptides, similar to recombinant expression. A more efficient route is to synthetically produce patellamides in a larger scale (Scheme 2) [35]. An additional benefit is that variations in terms of ligand structure, i.e. preorganization of the metal ion sites, as well as the donor sets and consequently the tuning of the electronics of the metal sites, are accessible.

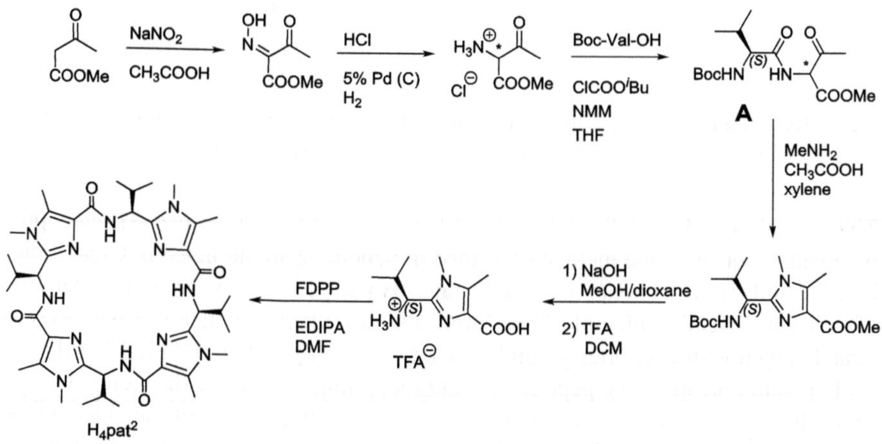

Scheme 2 Synthesis of H_4pat^2 [35]

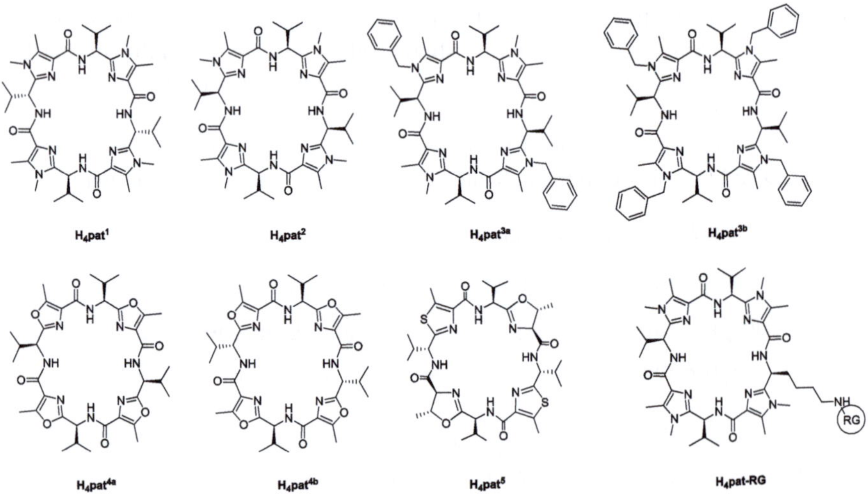

Chart 2 Synthetic *pseudo*-octapeptides H_4pat^1–H_4pat^5 and H_4pat-RG (RG: reporter group)

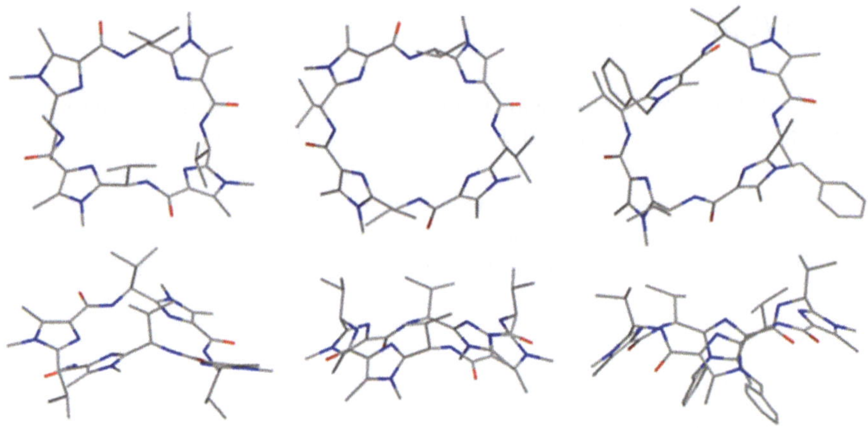

Fig. 3 Top and side views of the crystal structures of H_4pat^1 [43], H_4pat^2 [44], and H_4pat^{3a} (*grey*: C, *blue*: N, *red*: O, hydrogen atoms are omitted for clarity) [42]. Adapted with permission from [34]

The natural peptide ascidiacyclamide and four model ligands H_4pat^1–H_4pat^{3b}, prepared analogously to the method described in Scheme 2 are depicted in Chart 2 and Fig. 3. Similar routes to oxazole-based ligands H_4pat^{4a}, H_4pat^{4b}, a very close analogue of the natural ascidiacyclamide, H_4pat^5 were also described as well as a H_4pat^2 derived ligand tagged with a reporter group H_4pat-RG, all shown in Chart 2 [1, 34, 38].

Two components of the peptides are of crucial importance for the stereochemistry of the ligands: the heterocyclic moiety and the configuration of the side chains. The model patellamides H_4pat^1, H_4pat^{4b} and H_4pat^5 show an alternate *R,S,R,S* configuration as observed in the natural products, resulting in a saddle-shaped preorganized geometry of the ligand (see Scheme 1 and Fig. 3) [38]. H_4pat^2, H_4pat^{3a}, H_4pat^{3b} and H_4pat^{4a}, on the contrary, have an all-*S* configuration of the side chains and are therefore easier to prepare. They also lack, as does H_4pat^1 and H_4pat^{4b}, the two partially saturated heterocycles of the natural peptides. Only H_4pat^5 comprises both features, exhibiting oxazolines and thiazoles as heterocyclic moieties and the natural *R,S,R,S* configuration of the side chains. Apart from the side chain configurations, the heterocycles are important since the oxazoline is not planar and more flexible, and is therefore influencing the coordination geometry and metal ion binding behavior. Also, the different pK_a values of the heterocycles: *N*-methyl imidazole ≈7.0, oxazole 0.8, oxazoline 4.8, and thiazole 2.5, influence the binding behavior of the corresponding patellamide ligands [45]. How these three features separately and in combination influence the structural, electronic and kinetic properties of the Cu^{II} complexes is not yet clear in detail—a few features will be discussed at various parts of this chapter.

Cu^{II} and Zn^{II} Coordination Chemistry of Westiellamide and Patellamide Derivatives: Solution Equilibria and Structures

Studies on patellamide coordination chemistry have primarily concentrated on Cu^{II}. An important reason, apart from the alleged relevance of Cu^{II} for the ascidians [38], is that UV-vis-NIR, CD and EPR spectroscopy in combination with force field and/or DFT calculations have allowed a thorough analysis of the solution structures, equilibria and hydrolase-type reaction pathways of these complexes [34, 43, 45–50]. In particular, EPR spectroscopy in combination with structural modeling (MM-EPR, MM-DFT-EPR) has been developed as an important method for the determination of solution structures, in particular for dinuclear Cu^{II} complexes, where the interaction is dominated by dipole-dipole coupling [51–53]. A typical Spin Hamiltonian for dinuclear systems is given in Eq. (2). This describes the isotropic and anisotropic coupling of the individual Spin Hamiltonians of each of the two Cu^{II} centers *i* and *j*, as defined in Eq. (1), and a number of additional interactions, not discussed here in detail, have been included in Graeme Hanson's Sophe suite of programs (originally based on the DISSIM software developed by Smith and Pilbrow) [34, 54–57].

$$H = \beta_e B \cdot g \cdot S + S \cdot A({}^{63,65}Cu) \cdot I({}^{63,65}Cu) - g_n \beta_n B \cdot I$$

$$+ \sum_{3}^{j=1} (S_j \cdot A_j({}^{14,15}N) \cdot I_j({}^{14,15}N) - g_n \beta_n B_j \cdot I_j({}^{14,15}N)) \quad (1)$$

$$H = \sum_{2}^{i,j=l;l \neq j} H_i + J_{iso,ij} S_i \cdot S_j + S_i \cdot J_{ij} \cdot S_j \quad (2)$$

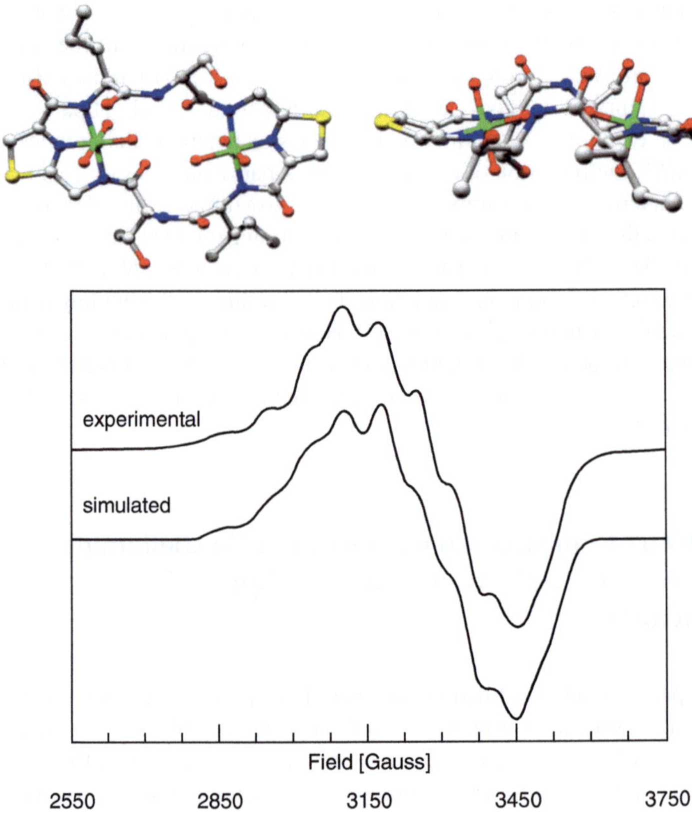

Fig. 4 Top and side views of the MM-EPR fitted structure of [Cu^{II}_2(patN)] (*grey*: C, *blue*: N, *red*: O, *green*: Cu, hydrogen atoms are omitted for clarity), and EPR spectrum (experimental, 9.594 GHz, and simulated) of the dicopper(II) complex (parameters for the simulation are given in Table 2) [47]. Adapted with permission from [47]

The first structure of a Cu^{II} complex of a synthetic patellamide derivative, and only the second after that of the carbonato-bridged Cu^{II} complex of ascidiacyclamide (see Fig. 2) was the solution structure of [Cu^{II}_2(patN)] with a synthetic patellamide derivative with two thiazole and two hydrolyzed oxazoline rings, determined with MM-EPR (see Fig. 4) [46, 47]. How important solution structures are, also in terms of solution equilibria and reactivities, emerges from Fig. 5 which shows two structures of carbonato-bridged dicopper(II) complexes. Switching between these two structures may be necessary to prevent product inhibition in the putative carbonic anhydrase catalytic cycle [34, 38, 58]. Dynamic equilibria between various possible structures (*e.g.* those in Fig. 4) should emerge from EPR spectroscopy but until recently, there were no examples where this has been observed and thoroughly analyzed [59].

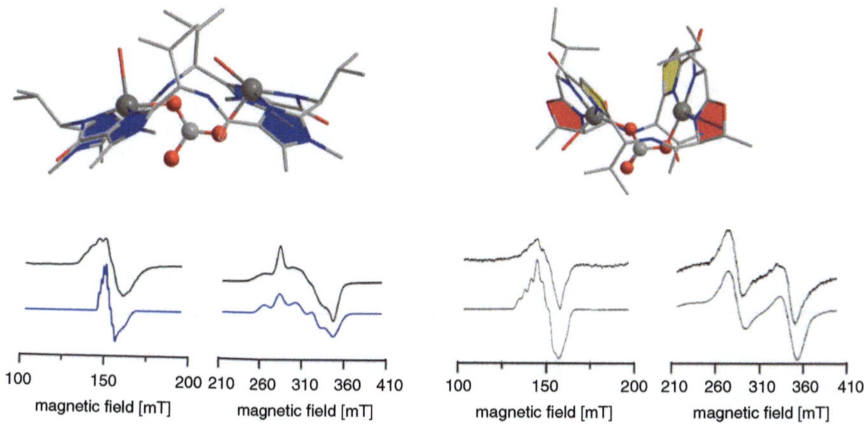

Fig. 5 Experimental (*black*) and simulated spectra (*blue*) of the carbonato bridged dinuclear Cu^{II} complex with the patellamide ligand H_4pat^2 (*left*) (9.3769 GHz) [34] and the natural peptide ascidiacyclamide (*right*) (9.4721 GHz) [41]. Reprinted and adapted with permission from [34] and [45]

Table 2 g and A- tensor values [10^{-4} cm^{-1}] and structural parameters (Cu•••Cu distances and Euler angles) of the dicopper(II) complex [Cu^{II}_2(patN)] given in Fig. 4

Fitted electronic field and structural parameters			
g_z	2.260	χ [°]	88
$g_{x,y}$	2.085	ρ [°]	25
A_z [10^{-4} cm^{-1}]	82	τ [°]	35
$A_{x,y}$ [10^{-4} cm^{-1}]	77	r_{Cu-Cu} [Å]	5.89

Cu^{II} complexes of the smaller *pseudo*-hexapeptide macrcocyles (westiellamide derivatives) have been studied in detail with combinations of CW-EPR, pulsed EPR, MCD and DFT calculations (Fig. 6) [45, 60]. Various mono- and dinuclear structures were observed, depending on the heterocycle and solvent. In addition, the combination of ENDOR, HYSCORE and MCD spectroscopy together with DFT calculations allowed to relate subtle structural differences with the spin distribution, and this clearly has relevance to specific reactivities.

Recent investigations demonstrate that the model ligands are complementary for the binding of Cu^{II} and Zn^{II}, like reported for the natural patellamides. Isothermal calorimetric studies indicate a higher stability constant for the dinuclear Cu^{II} complex with ligand H_4pat^1, a ligand composed of building blocks with alternating *R* and *S* side chains, mimicking the side chain arrangement of the natural patellamides ($\log K$ = 6.2, methanolic solution) [38]. The Cu^{II} complexes of ligands H_4pat^2, H_4pat^{3a} and H_4pat^{4a} show lower stabilities: $\log K = 4.6; 5.2; 5.5$; which is in the range of the stability constants observed for the Cu^{II} binding of natural octapeptides [61]. From the collective data on the Cu^{II} binding behavior of the model ligands, a cooperative binding is implicated [38].

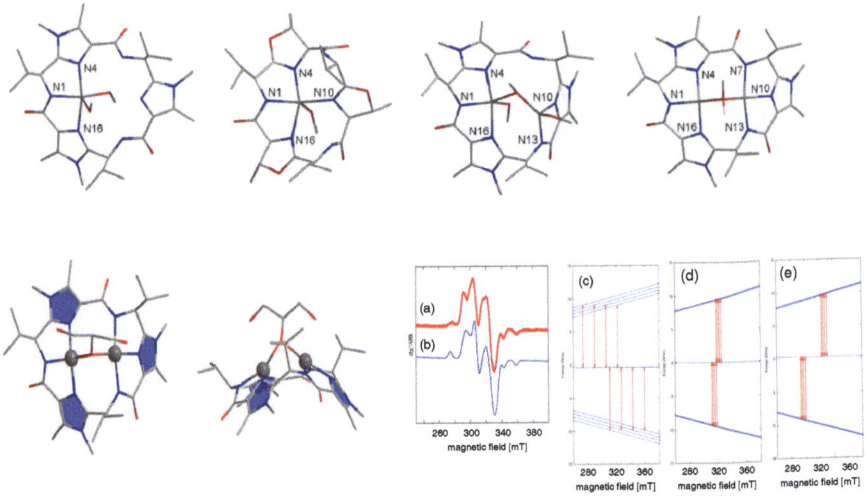

Fig. 6 *Top*: two different structures each for mono- and dinuclear Cu^{II} complexes of westiellamide derivatives (cyclic *pseudo*-hexapeptides with imidazole and oxazole heterocycles (*blue*: N, *red*: O). Bottom left: DFT-calculated structure (Gaussian03, B3LYP/6-31g*/TZVP) of the glycerol-bridged dinuclear Cu^{II} complex with the imidazole-based cyclic peptide (compare the computed structure *top right* with a bridging methanol). *Bottom right*: X-Band EPR spectra of the glycerol-bridged dicopper(II) complex. (**a**) Experimental EPR spectrum in a methanol/glycerol (1:1), frozen solution, $c(H_3L^1) = 1.5$ mM, $\nu = 9.434401$ GHz, T = 50 K; (**b**) computer simulation of (**a**); (**c–e**) energy level diagrams showing the allowed EPR transitions along the z, x and y principal directions, respectively

UV-vis-, CD-, EPR-spectroscopy and mass spectrometry studies of the Cu^{II} complexes with various model patellamides [34] revealed the solution equilibria shown in Scheme 3 (*n.b.* that the outside coordination does only occur for oxazole-based model patellamides). All investigations presented here were carried out in methanol with MeO^- as the base for solubility reasons. Therefore, the pH of the solutions under investigation is not known and the results must hence be interpreted carefully, when speaking about the function and structure of the complexes at aqueous conditions. Scheme 3 shows two different arrangements of the dinuclear Cu^{II} patellamide complex with one hydroxido-coligand: (a) the terminal binding mode and (b) the bridging mode, the latter of which has been confirmed with X-ray crystal structure determination (see Fig. 7) [38].

The binding motif of Cu^{II} in $[Cu_2(H_2pat^1)(\mu\text{-}OH)(H_2O)_2]^+$ is in agreement with expectations from the crystal structure of the dinuclear Cu^{II} complex of ascidiacyclamide—the Cu^{II} cations are coordinated by two heterocyclic nitrogen atoms (imidazole in this case), and a deprotonated amide nitrogen. Additionally, each of the Cu^{II} ions is coordinated by a water molecule and a bridging hydroxide coligand. This results in a square-pyramidal coordination environment for the Cu^{II} ions, which is also expected from EPR spectroscopy and agrees with the predictions from DFT calculations [34, 62].

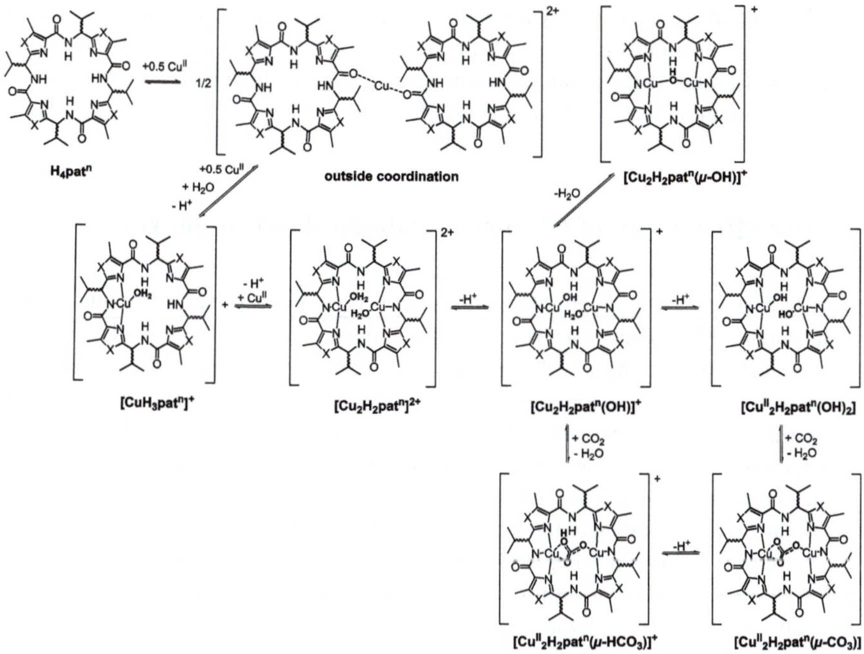

Scheme 3 Cu^{II} complexation equilibria of model patellamides H_4pat^n [34]

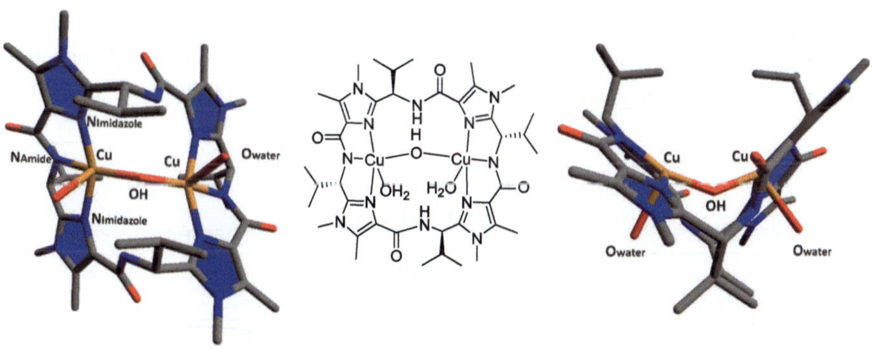

Fig. 7 Top and side view of $[Cu_2(H_2pat\ [1])(\mu\text{-}OH)(H_2O)_2]^+$ (*orange*: Cu, *grey*: C, *blue*: N, *red*: O, hydrogen atoms are omitted for clarity) [38]

Studies on Zn^{II} complexes with model patellamides have been recently carried out, indicating an overall similar coordination behavior as observed for the respective Cu^{II} species; NMR spectroscopy in combination with EPR spectroscopy, in particular of the heterodinuclear complexes, was instrumental for the structural assignments [49]. Isothermal calorimetric measurements give an ambiguous number of binding events, i.e. 1.51 ± 0.16 (H_4pat^1) and 1.69 ± 0.78 (H_4pat^2); however results from mass spectrometry investigations indicate the complexes to be dinuclear. The stability constants determined for these species are $\log K_1K_2 = 5.0$ and 4.5 for H_4pat^1 and H_4pat^2 (methanolic solution), respectively, which is in accordance with studies on natural peptides (*vide supra*) [49].

In vitro Reactivity of Dinuclear Patellamide Complexes

The driving force of patellamide complex investigations was the urge to understand their potential biological role. As pointed out above, the slightly basic pH of seawater supports the assumption that the Cu^{II} patellamide complexes are formed in *L. patella* and *Prochloron*. Since the determined binding constants with Cu^{II} and Zn^{II} are rather small, the function of the patellamides as chelators performing metal ion storage and transport is not very likely. Consequently, the potential role of the dinuclear complexes as catalysts was the focus of research. Dinuclear metal ion sites have various advantages compared to just one metal ion in the active site, i.e. a lower thermodynamic driving force for redox reactions due to charge-delocalization, a decreased activation barrier for solvent and enzyme reorganization, the preorientation and electrostatic activation of substrates as well as the easier formation of nucleophiles, which could in turn initiate hydrolysis [63]. Moreover, the dinuclear sites are capable of stabilizing transition states that are involved in the hydrolysis [63]. Since dinuclear metal ion complexes are abundant as effective catalysts in nature, several studies on the putative catalytic reactivity of the model patellamide complexes were carried out.

Carbonic anhydrase reactivity was the first to be demonstrated by the dicopper(II) complexes of synthetic patellamide derivatives (see Scheme 4 for the putative catalytic

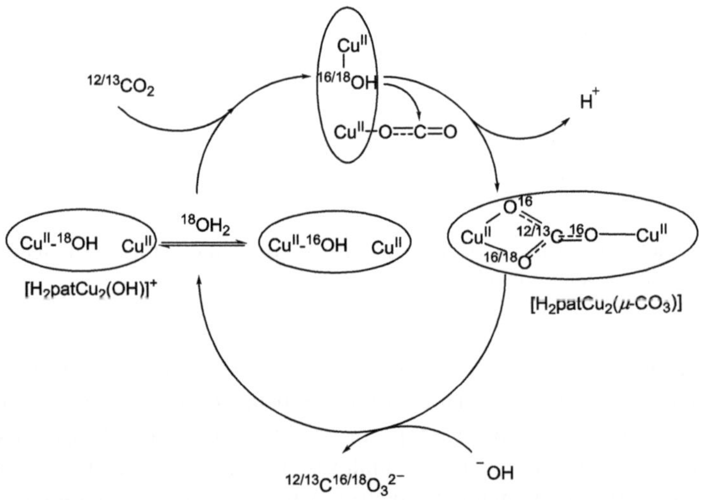

Scheme 4 Schematic representation of expected carbonic anhydrase-like activity of the dinuclear Cu^{II} patellamide complexes, adapted with permission from reference [64].

cycle and Fig. 5 for the relevant structures. The in vitro studies demonstrate that these complexes are the most efficient functional models for carbonic anhydrase and—if this was the biological function, the patellamide based catalysts would be the first Cu^{II}-based carbonic anhydrases. The study also shows that the efficiency strongly depends on the configuration of the side-chains, and the natural *R,S,R,S* configuration is needed for an efficient catalytic process [38].

The discovery of carbonic anhydrase activity suggested that the dicopper(II) complexes of the cyclic *pseudo*-octapeptides might be capable of forming hydroxide ions, allowing the activity as a hydrolase-like catalyst. Phosphatase activity therefore was another candidate for the putative biological function of these metabolites. A first set of experiments strongly indicated that this is indeed the case [65], and a full study involving the entire set of synthetic patellamide derivatives and dicopper(II) as well as dizinc(II) and mixed Zn^{II}/Cu^{II} complexes shows that model phosphoester substrates (mono- and diesters) are efficiently hydrolyzed [49]. From the slightly lower concentration of Zn^{II} and the slightly smaller complex stabilities in comparison with Cu^{II} as well as from the relative catalytic efficiencies, it appears that the dicopper(II) complexes are likely to be the active phosphatase catalysts—if this should be the biological function [49].

The dinuclear Cu^{II} complexes of patellamide model ligands were also tested for glycosidase- and lactamase-like activities [50]. Interestingly, the Cu^{II} complexes of the tested model ligands show efficient catalytic reactivity with model substrates for lactamase and glycosidase assays. However, while the optimum pH range for phosphatase- and carbonic anhydrase-like activities of the dinuclear patellamide-based catalysts is around pH 7–8, those for the glycosidase- and lactamase-like reactivities are around pH 10–11 [49, 50, 58, 65]. Importantly, a recent study showed that the pH in *L. patella* varies significantly from pH 7 in the darkness to around pH 11 in the light [4]. Therefore, the patellamide complexes might be light sensitive and therefore pH-dependent catalysts.

Preliminary In vivo Studies of Patellamide Complexes

The concentration of copper cations in living cells and their oxidation state is a topic that is not well studied for cyanobacteria. It is known that mammalian (eukaryotic) cells exhibit a reductive interior atmosphere [66], and intracellular copper cations not bound to biomolecules are thus present as Cu^{I} ions [66]. The copper ion homeostasis is regulated by a copper ion importing and exporting system and leads to a Cu cation concentration in the cell in the femtomolar range [66]. Similar systems were also found for the prokaryotes *Enterobacteria* [67]. Here, interior Cu^{II} concentrations of 10^{-6} M were reported, as compared to 10^{-8} M in the surrounding medium [67]. However, no data are available for *Prochloron* and hence one can only speculate the concentrations to be similar to the prokaryotes mentioned before.

In addition, the question about the reductive/oxidative milieu in the cells remains. What is known is that, depending on the level of irradiance, the pH is fluctuating and

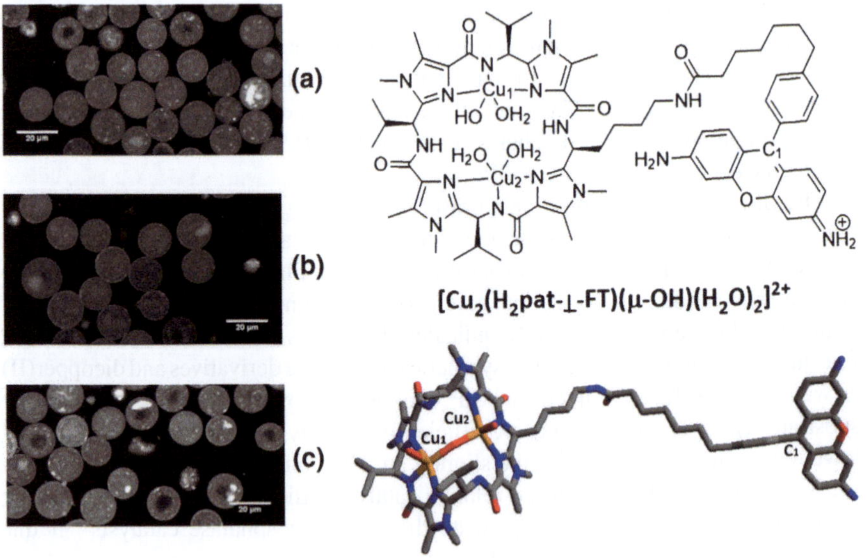

Fig. 8 *Left*: Confocal microscopy pictures of *Prochloron* cells treated (**a**) with H_4pat-Atto550, (**b**) with H_4pat-Atto550 and Cu^{II}, and (**c**) with H_4pat-Atto550 and Cu^{II} and subsequently with cyclam. *Right*: computed structure of the dinuclear Cu^{II} complexes of H_4pat-Atto550

the same is true for the level of oxygen saturation (*n.b.*: the oxygen saturation and pH were not measured inside *Prochloron* cells, but in close proximity to it) [4]. As a consequence of this oscillation during day/night time, the oxidative environment might be correlated to the level of oxygen in the cells, but so far there is no proof for this hypothesis. If this was the case, one would expect an alkaline environment (pH 11) as well as dioxygen super saturation upon solar irradiation [4] and consequently copper cations to be present as Cu^{II}. In the darkness, on the other hand, an interior milieu of pH 7.0 is expected together with an anoxic environment [4], which would favor Cu^{I}.

Although synthetic models of the natural patellamides bind Cu^{II} and exhibit catalytic activity in vitro as dicopper(II) complexes, their biological role is still not known. In order to study the question as to whether the complexes of the cyclic *pseudo*-octapeptides are also formed inside the *Prochloron* cells a patellamide ligand with an appended fluorescent group, H_4pat-Atto 550 was designed and studied concerning its binding behavior towards Cu^{II} in vivo (see Fig. 8).

The ligand H_4pat-Atto550 was initially studied subject to its in vitro binding behavior in TRIS buffer at pH 8.2. Quenching experiments were carried out and a Stern-Volmer constant of $K_s = 1.35(2) \times 10^6\ M^{-1}$ was deduced, which is within expectations for the stability constants of the respective dinuclear Cu^{II} complexes of H_4pat^1 and H_4pat^2 [1]. In a second step the uptake of the ligand with the appended fluorescent tag was investigated and it could be shown, that heat shock treatment (42 °C, 40 s) was sufficient to maintain the *Prochloron* cells photosynthetically active (proved by PAM measurements) and to stimulate the cells to take up the ligand [1]. Finally, the Cu^{II} ion complexation behavior of the ligands inside the cells was investigated. For that three different populations were subject to observation: (a) a population of *Prochloron* cells treated with H_4pat-Atto550, (b) a population of *Prochloron* cells treated with the ligand H_4pat-Atto550 and Cu^{II} and (c) a population that was treated with cyclam as a competitive chelator after the addition of Cu^{II} [1]. Figure 8 shows the respective confocal microscopy pictures of a sample of each of the populations [1]. Subsequent population analysis together with the results from flow cytometry experiments show a smaller relative fluorescence intensity of the fluorescent tag for population (b) as compared to (a) and (c). This is caused by the formation of a non-fluorescent species upon addition of Cu^{II}. This implies that the dinuclear Cu^{II} complexes are stable in *Prochloron* cells.

In order to verify this finding, an additional technique should be applied to unambiguously elucidate the stability of the Cu^{II} complexes in the cells. Here, EPR spectroscopy would be a helpful tool, for example by the investigation of electron spin - nuclear quadrupole interactions of a specifically deuterated patellamide with Cu^{II}. This could be achieved by ESEEM or HYSCORE measurements, which exploit the modulation of the exponential decay of the electron spin echo caused by the interaction of the electron spin with the quadrupole moment of nuclei in proximity to the paramagnetic center [68, 69]. The application of such pulsed EPR techniques would allow the explicit study of patellamide-Cu^{II} interactions and lead to a minor background from other Cu^{II} proteins. Consequently, a deuterated patellamide is currently being prepared to apply this technique. Once the in vitro EPR experiments will be brought to a stage that allows the validation of the binding of Cu^{II} to the patellamides, in vivo experiments with *Prochloron* will have to be carried out. Therefore, the ligand will have to be brought into *Prochloron* cells, as studied in detail for the H_4pat-Atto550 ligand, discussed above. Information from these studies will provide a more detailed understanding of the binding site of Cu^{II} to the patellamides in *Prochloron* cells and will validate the results obtained from the fluorescence quenching experiments.

Conclusion and Outlook

The cyclic *pseudo*-octapeptides of the ascidians have drawn attention of marine biochemists, medicinal, organic and coordination chemists for about three decades. Their Cu^{II} and to a lesser content Zn^{II} coordination chemistry is rather well understood. A range of in vitro studies have helped to discover various possible biological functions of these metabolites and their metal complexes. First in vivo studies have now been carried out suggesting that Cu^{II} complexes of patellamide derivatives might be present in cells. It is quite obvious that well-designed in vivo studies which should involve EPR spectroscopy are now needed to shed light on the structures and reactivities of these complexes in the *Prochloron* cells.

Acknowledgements We are grateful for the excellent work done by coworkers, collaborators and colleagues mentioned in the references, specifically the work done by PhD students in the Comba group and the collaboration with colleagues from Brisbane, particularly Lawrie Gahan and Graeme Hanson. Graeme's input was invaluable and we miss him as a scholar and dear friend.

References

1. Eisenschmidt A (2016) Dinuclear copper(II) patellamide complexes: studies on their potential hydrolase-like activity and in vivo stabilities, PhD thesis. Heidelberg University, Heidelberg
2. Mayer H (2016) Frankfurter Allgemeine Zeitung (Feuilleton) 31:08
3. Kaim W, Schwederski B (2005) Bioanorganische Chemie—Zur Funktion chemischer Elemente in Lebensprozessen, 4th edn. Teubner, Wiesbaden
4. Kühl M, Behrendt L, Trampe E, Qvortrup K, Schreiber U, Borisov SM, Klimant I, Larkum AW (2012) Front Microbiol 3:402
5. Mereschkowsky K (1910) Biol Centralbl 30:353
6. Schwartz W (1973) Z Allg Mikrobiol 13:186
7. Whatley JM, John P, Whatley FR (1979) Proc Roy Soc London Series B Biol Sc 204:165
8. Lewin RA, Cheng L (1989) Prochloron: a microbial enigma. Chapman and Hall, London
9. Lewin RA, Withers NW (1975) Nature 256:735
10. Newcomb EH, Pugh TD (1975) Nature 235:533
11. Hirose E, Neilan BA, Schmidt EW, Murakami A (2009) A enigmatic life and evolution of prochloron and realted cyanobacteria inhabitin colonial ascidians. Nova Science Publishers, Inc., Hauppauge, New York
12. Donia MS, Fricke WF, Partensky F, Cox J, Elshahawi SI, White JR, Phillippy AM, Schatz MC, Piel J, Haygood MG, Ravel J, Schmidt EW (2011) Proc Natl Acad Sci 108:E1423
13. Griffiths DJ (2006) Bot Rev 72:330
14. Odintsov VS (1990) Endocytobiosis Cell Res 7:253
15. Donia MS, Fricke WF, Ravel J, Schmidt EW (2011) PLoS One 6:17897
16. Behrendt L, Larkum AW, Trampe E, Norman A, Sorensen SJ, Kühl M (2012) ISME J 6:1222
17. Giordano M, Beardall J, Raven JA (2005) Annu Rev Plant Biol 56:99
18. Critchley C, Andrews TJ (1984) Arch Microbiol 138:247
19. Schmidt EW, Donia MS (2010) Curr Opin Biotechnol 21:827
20. Schmidt EW, Donia MS, Mcintosh JA, Fricke WF, Ravel J (2012) J Nat Prod 75:295
21. Pettit GR, Day JF, Hartwell JL, Wood HB (1970) Nature 227:962
22. Krebs HC (1986) Progress in the chemistry of organic natural products. Springer, Wien, pp 151–363
23. Milne BF, Long PF, Starcevic A, Hranueli D, Jaspars M (2006) Org Biomol Chem 4:631–638
24. Schmidt EW, Nelson JT, Rasko DA, Sudek S, Eisen JA, Haygood MG, Ravel J (2005) Proc Natl Acad Sci 102:7315–7320
25. Houssen WE, Bent AF, McEwan AR, Pieiller N, Tabudravu J, Koehnke J, Mann G, Adaba RI, Thomas L, Hawas UW, Liu H, Schwarz-Linek U, Smith MCM, Naismith JH, Jaspars M (2014) Angew Chem Int Ed 53:14171–14174
26. Long PF, Dunlap WC, Battershill CN, Jaspars M (2005) ChemBioChem 6:1760
27. Houssen WE, Jaspars M (2010) ChemBioChem 11:1803
28. Hamamoto Y, Endo M, Nakagawa T, Nakanishi T, Mizukawa K (1983) J Chem Soc Chem Commun 1983:323–324
29. Schmidt U, Utz R, Gleich P (1985) Tetrahedron Lett 26(36):4367–4370
30. Ireland CS, Paul J (1980) J Am Chem Soc 102(17):5688–5691
31. Prinsep MRM, Moore RE, Levine IA, Patterson GML (1992) J Nat Prod 55(1):140–142
32. Hambley TW, Hawkins CJ, Lavin MF, van den Brenk A, Watters DJ (1992) Tetrahedron 48(2):341–348

33. Haberhauer G (2008) Tetrahedron Lett 49:2421
34. Comba P, Dovalil N, Gahan LR, Haberhauer G, Hanson GR, Noble CJ, Seibold B, Vadivelu P (2012) Chem Eur J 18:2578–2590
35. Haberhauer G, Rominger F (2003) Eur J Org Chem 2003:3209–3218
36. Haberhauer G, Drosdow E, Oeser T, Rominger F (2008) Tetrahedron 64:1853–1859
37. Pintér Á, Haberhauer G (2009) Tetrahedron 65:2217
38. Comba P, Dovalil N, Gahan LR, Hanson GR, Westphal M (2014) Dalton Trans (Dalton Perspective) 43:1935–1956
39. Wipf P, Fritch PC, Geib SJ, Sefler AM (1998) J Am Chem Soc 120:4105–4112
40. Degnan BM, Hawkins CJ, Lavin MF, Mccaffrey EJ, Parry DL, van den Brenk AL, Watters DJ (1989) J Med Chem 32(6):1349–1354
41. van den Brenk AL, Byriel KA, Fairlie DP, Gahan LR, Hanson GR, Hawkins CJ, Jones A, Kennard CHL, Moubaraki B, Murray KS (1994) Inorg Chem 33:3549–3557
42. Comba P, Dovalil N, Haberhauer G, Kowski K, Mehrkens N, Westphal M (2013) Z Anorg Allg Chem 639(8-9):1395–1400
43. Comba P, Dovalil N, Hanson GR, Linti G (2011) Inorg Chem 50:5165–5174
44. Haberhauer G, Pinter A, Oeser T, Rominger F (2007) Eur J Org Chem 2007:1779–1792
45. Comba P, Gahan LR, Haberhauer G, Hanson GA, Noble CJ, Seibold B, van den Brenk AL (2008) Chem Eur J 14:4393–4403
46. Bernhardt PV, Comba P, Hambley TW, Massoud SS, Stebler S (1992) Inorg Chem 31:2644–2651
47. Comba P, Cusack R, Fairlie DP, Gahan LR, Hanson GR, Kazmaier U, Ramlow A (1998) Inorg Chem 37:6721–6727
48. Bernhardt PV, Comba P, Fairlie DP, Gahan LA, Hanson GR, Lötzbeyer L (2002) Chem Eur J 8:1527–1536
49. Comba P, Eisenschmidt A, Gahan L, Hanson G, Mehrkens N, Westphal M (2016) Dalton Trans 45:18931–18945. doi:10.1039/C6DT03787A
50. Comba P, Eisenschmidt A, Kipper N, Schiessl JJ (2016) Inorg Biochem 159:70–75
51. Comba P (1992) Habilitation. Universität Basel
52. Comba P, Lampeka YD, Lötzbeyer L, Prikhod'ko AI (2003) Eur J Inorg Chem 2003:34–37
53. Comba P, Lampeka YD, Prikhod'ko AI, Rajaraman G (2006) Inorg Chem 45:3632–3638
54. Smith TD, Pilbrow JR (1974) Coord Chem Rev 13:173–278
55. Comba P, Pandian S, Wadepohl H, Wiesner S (2011) Inorg Chim Acta (Special Issue Wolfgang Kaim) 374:422–428
56. Hanson GR, Gates KE, Noble CJ, Griffin M, Mitchell A, Benson S (2004) J Inorg Biochem 98:903–916
57. Hanson GR, Noble CJ, Benson S (2009) Molecular sophe, an integrated approach to the structural characterization of metalloproteins - the next generation of computer simulation software. In: Hanson GR, Berliner LJ (eds) High resolution EPR: applications to metalloenzymes and metals in medicine, vol 28. Springer, Dordrecht, Heidelberg, London, New York, pp 105–174
58. Comba P, Gahan LR, Hanson GR, Maeder M, Westphal M (2014) Dalton Trans 43(8):3144–3152
59. Bosch S, Comba P, Gahan LR, Hanson GR, Schenk G, Wadepohl H (2015) Chem Eur J 21:18269–18279
60. Comba P, Dovalil N, Hanson G, Harmer J, Noble CJ, Riley MJ, Seibold B (2014) Inorg Chem 53:12323–12336
61. Comba P, Dovalil N, Haberhauer G, Kowski K, Mehrkens N, Westphal MZ (2013) Allg Anorg Chem (Special Issue Bioinorg Chem) 639:1395–1400
62. Dovalil N (2010) Copper(II) coordination chemistry of cyclic pseudo hexa- and octapeptides, PhD thesis, Heidelberg University, Heidelberg
63. Daumann L (2014) Dalton Trans 43:910
64. Comba P, Dovalil N, Gahan LR, Haberhauer G, Hanson GR, Noble CJ, Seibold B, Vadivelu P (2012) Chemistry 18(9):2578–2590. doi:10.1002/chem.201101975.

65. Comba P, Gahan LR, Hanson GR, Westphal M (2012) Chem Comm 48(75):9364. doi:10.1039/c2cc34836e
66. Foster AW, Osman D, Robinson NJJ (2014) Biol Chem 289(41):28095–28103. doi:10.1074/jbc.R114.588145
67. Porcheron G, Garenaux A, Proulx J, Sabri M, Dozois C (2013) FCIMB 3(90). doi:10.3389/fcimb2013.00090
68. Rowan LG, Hahn EL, Mims WB (1965) Phys Rev A 137(1A):61–71
69. Mims WB (1972) Phys Rev B 5(7):2409–2419

Paramagnetic Resonance of High-Spin Co(II) in Biologically-Relevant Environments: Models to Metalloproteins

Robert R. Baum, Christopher D. James, and David L. Tierney

Abstract The use of Co(II) as a spectroscopic surrogate for Zn(II) is now a well-established protocol in metallobiochemistry. The d^7 Co(II) ion usually adopts a high-spin ($S = 3/2$) configuration and a coordination geometry similar to the native Zn(II) ion, often returning an active enzyme. However, the complicated electronic structure that gives rise to easily detectable signals in a wide array of optical and magnetic spectroscopies simultaneously hampers data interpretation in terms of structure. Nowhere is this more evident than in the EPR spectra of Co(II) complexes, particularly at X-band. Some alternatives to common practice in the assignment and simulation of high-spin Co(II) EPR are presented. Our intent is to shed light on the sources of spectral complexity, and address some of the remaining issues confounding the successful application of more advanced techniques, such as ENDOR (CW or pulsed) and ESEEM. The importance of spin sub-level mixing into the magnetic ground state, leading to possible intensity stealing and the appearance of signals from both spin doublets, is discussed in terms of available zero-field splitting data, and the identity of the ground doublet.

Keywords Cobalt • High-spin • Electronic structure • Zero-field splitting • Pyrazolylborate • Parallel mode EPR • Pulsed EPR • ENDOR • ESEEM

Introduction

The use of Co(II) as a spectroscopic surrogate for Zn(II) was first reported in 1958, and is now a well-established protocol in metallobiochemistry [1, 2]. The d^7 Co(II) ion usually adopts a high-spin (*hs*, $S = {}^3/_2$) configuration and a coordination geometry similar to the native Zn(II) ion, often returning an active enzyme. Unlike Zn(II), which is only accessible by X-ray absorption spectroscopy (XAS) [3], Co(II) is amenable to a wide array of techniques, including optical spectroscopy, magnetic

R.R. Baum • C.D. James • D.L. Tierney (✉)
Department of Chemistry and Biochemistry, Miami University, Oxford, OH 45056, USA
e-mail: tiernedl@miamioh.edu

G. Hanson, L. Berliner (eds.), *Future Directions in Metalloprotein and Metalloenzyme Research*, Biological Magnetic Resonance 33,
DOI 10.1007/978-3-319-59100-1_3

circular dichroism (MCD) [4], XAS [3], EPR [5], NMR [6] and, more recently, electron-nuclear double resonance (ENDOR) [7–9] and high-frequency/field EPR (HF/F-EPR) [10–15]. While enabling its utility as a spectroscopic probe, the complex electronic structure of *hs* Co(II) also facilitates its widespread use in homogeneous [16, 17] and heterogeneous catalysts [18–20], magnetic materials [21, 22] and single-molecule magnets [23, 24]. However, the complicated electronic structure that gives rise to easily detectable signals in all of the above techniques, simultaneously hampers the interpretation of spectroscopic signatures in terms of structure. The presence of unquenched orbital momentum gives rise to thermally accessible excited states, and allows for excited state character to potentially mix into the electronic/magnetic ground state.

The complexities of *hs* Co(II) EPR spectra at X-band are notorious (see Fig. 1), and the EPR literature associated with *hs* Co(II) small molecules is vast (~3000 reports) and impossible to characterize. EPR of Co(II)-substituted proteins was reviewed in this series in 2010 [5], focusing heavily on the practicalities of using Co(II) as an EPR probe, including methods of incorporation, proper choice of experimental conditions and interpretation of spectra in terms of structure. Our goal with the present review, focused on the properties of small molecule model compounds, was to step back from metalloproteins, discuss the electronic structure in more detail, and then discuss what we have learned on the basis of our small molecule studies. We will present some alternatives to common practice in the assignment and simulation of *hs* Co(II) EPR. We hope to shed light on the sources of the complexity, and address some of the remaining issues confounding the successful application of more advanced techniques, such as ENDOR (CW or pulsed) and ESEEM. To do this, it is imperative that we understand the nature of the magnetic ground state, and the importance of spin sub-level mixing.

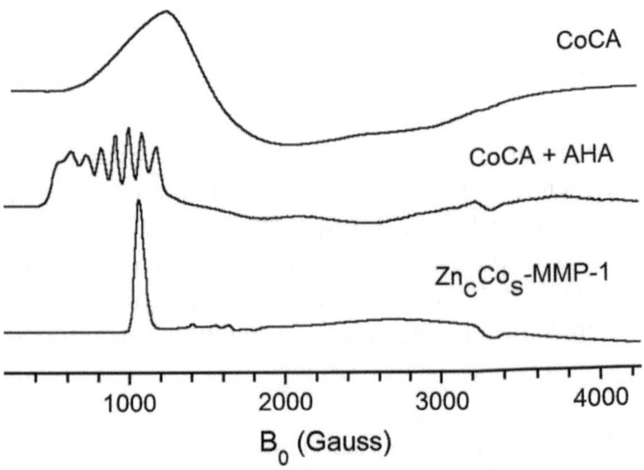

Fig. 1 Examples of low-temperature X-band EPR of Co(II)-substituted carbonic anhydrase (CoCA, *top*), CoCA + acetohydroxamate (AHA) and Co(II) in the structural site of matrix metalloproteinase 1 (MMP-1)

Electronic Structure of High-Spin Co(II)

A review of the electronic structure that gives rise to the Co(II) spin states probed by EPR is useful here [25–31]. In the absence of a ligand field, the ^{4}F free-ion ground state ($L = 3$, $S = ^3/_2$, Fig. 2a is 28-fold degenerate, due to seven orbitally degenerate levels ($M_L = \pm3, \pm2, \pm1, 0$), each four-fold degenerate in spin ($M_S = \pm^3/_2, \pm^1/_2$), as illustrated in the abridged microstate table of Fig. 2b. The lowest energy quartet excited term is ^{4}P; however, it is important to note that the two lowest energy doublet states (^{2}H and ^{2}P), which are degenerate, are accidentally degenerate with the ^{4}P at C/B ~ 3.8 (where B and C are the Racah parameters) [26]. This value is very close to the free ion value of 4.3, while the value most commonly used in AOM calculations (4.6) places the doublet states slightly below the ^{4}P. Under these circumstances, the doublet states are capable of mixing into the ^{4}P excited state, and it is these spin-forbidden quartet-doublet transitions that give rise to multiplet structure in the visible absorption band of six-coordinate Co(II) complexes [29]. Their importance to the discussion below arises from the potential for the $^4T_1$(P) levels to mix with the $^4T_1$(F) levels. With typical values of B ranging from ~700 to 900 cm^{-1}, an energy of 15B places the ^{4}P about 10,000–12,000 cm^{-1} above the ground state, compared to crystal field splittings (Δ in Fig. 2) of ~3000–5000 cm^{-1} for tetrahedral complexes, and generally increasing with increasing coordination number [29, 32].

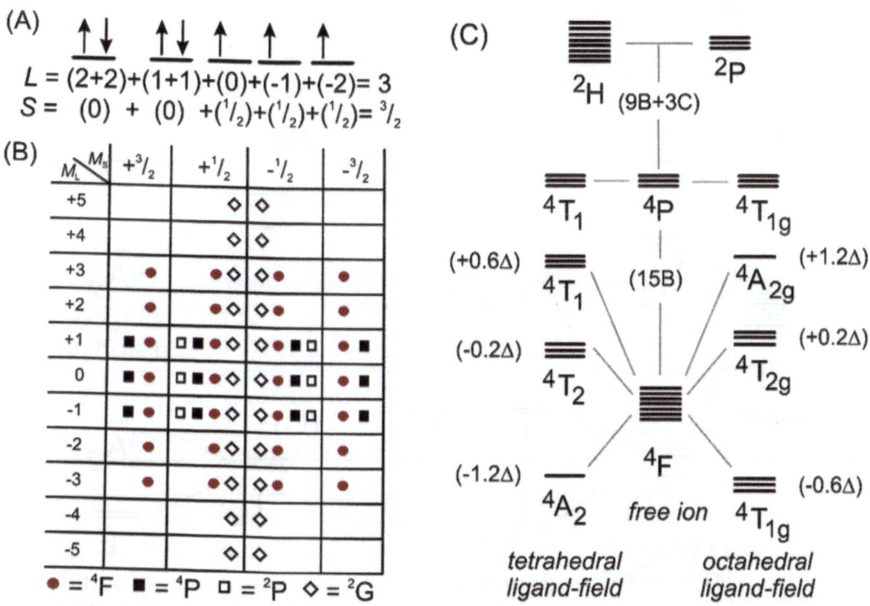

Fig. 2 (**a**) Ground state electronic structure of a d^7 ion in vacuum. (**b**) Microstate table for d^7, showing the four lowest electronic levels. Quartet states are shown with *solid symbols*; doublet states with *open symbols*. (**c**) Splitting of the d^7 free-ion states, depending on the crystal field splitting energy Δ, and the Racah B and C parameters

Distortions from spherical symmetry, such as those that accompany the introduction of a ligand field, partially lift the orbital degeneracy of the free ion, leading to two orbital triplets (T) and one orbital singlet (A). Their relative disposition in energy is described by the well-known Orgel diagram [33], represented in Fig. 2c. In short, a tetrahedral ligand field ($\Delta < 0$, left side of the diagram) stabilizes the orbital singlet, while an octahedral ligand field stabilizes the orbital triplet ($\Delta > 0$, right side of the diagram).

When the orbital triplet is lowest (right side of Fig. 2c), the spin system is defined by spin-orbit coupling, which splits each orbital triplet into six Kramers doublets, as shown in Fig. 3a. The three multiplets correspond to $J = {}^1/_2$ ($M_J = \pm{}^1/_2$), $J = {}^3/_2$ ($M_J = \pm{}^1/_2, \pm{}^3/_2$) and $J = {}^5/_2$ ($M_J = \pm{}^1/_2, \pm{}^3/_2, \pm{}^5/_2$), with energies given by $\frac{1}{2}\lambda\left[J(J+1)-L(L+1)-S(S+1)\right]$. Hund's third rule places the largest J multiplet lowest in energy [31], with the free ion spin-orbit coupling constant defined as negative ($\lambda_{Co} = \zeta/2S = -175\ \text{cm}^{-1}$) [29]. Lower-symmetry structural distortions can remove some or all of the ground state orbital degeneracy, depending on whether the direction of the distortion is on-axis (tetragonal, Fig. 3b) or off-axis (trigonal, Fig. 3c). A tetragonal elongation leaves the orbital singlet ($^4A_2$) lowest in energy, and the axes of quantization are unchanged from the undistorted octahedron. Meanwhile, a trigonal elongation leaves the orbital doublet (^4E) lowest and re-quantizes the orbitals along the three-fold axis, resulting in a spin-orbit coupling defined ground state.

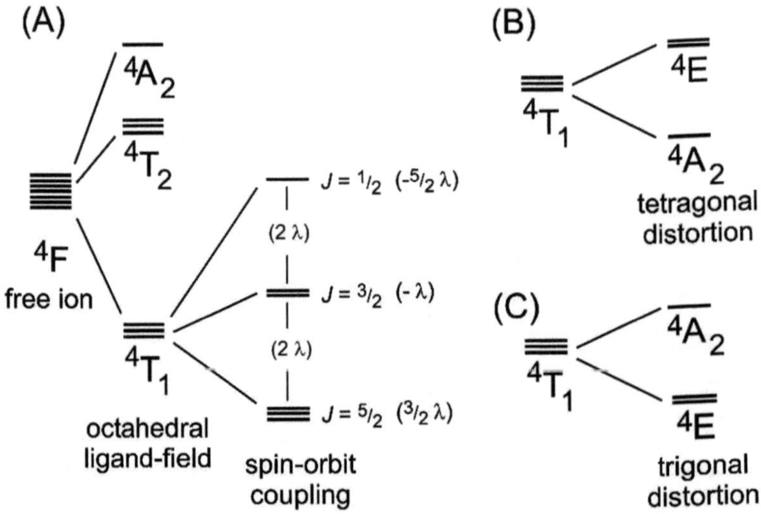

Fig. 3 (**a**) Splitting induced by spin-orbit coupling within the orbital triplet of an octahedral ligand field. The *lines* represent orbital levels, each itself is a spin quartet. (**b**) Tetragonal distortion. (**c**) Trigonal distortion

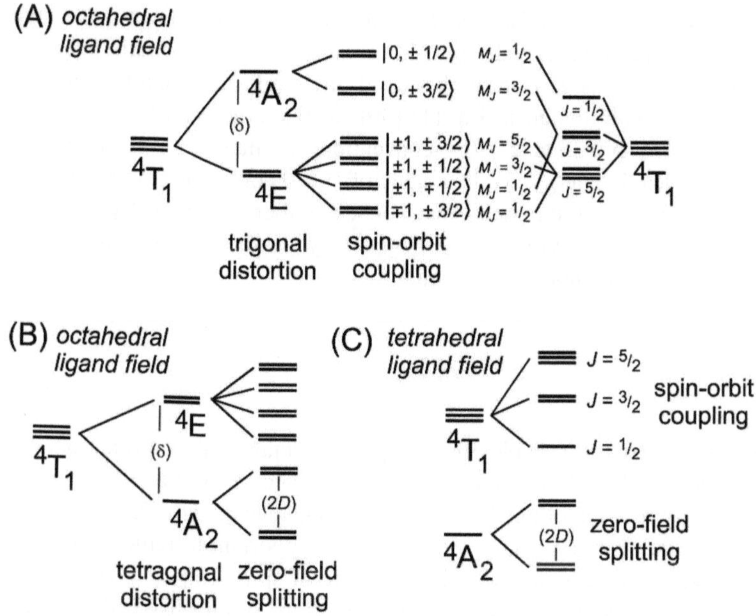

Fig. 4 (**a**) Effect of a trigonal distortion on the spin-orbit energy levels of an octahedral ligand field. The approximate $|M_L, M_S>$ wavefunctions, values of M_J and their origin in the spin-orbit basis, are shown to the *right*. (**b**) Effect of a tetragonal distortion. (**c**) Tetrahedral ligand field

The trigonally distorted case, shown in detail in Fig. 4a, has been well-described theoretically [26, 28, 30, 34]. The ground spin-orbit level, originating in the ^4E electronic level, is an $M_J = \pm^1/_2$ doublet, with wavefunctions that are linear combinations of the three $M_J = \pm^1/_2$ levels, $\Psi^{\pm 1/2} = a\left|-1,\frac{3}{2}\right\rangle + b\left|0,\frac{1}{2}\right\rangle + c\left|1,-\frac{1}{2}\right\rangle$. In the most extensively studied example (below) [8, 9, 34, 35], the trigonal splitting is sufficient that the ground doublet is ~95% $|-1,\frac{3}{2}>$, although it is important to note that even in this example, the ground doublet contains ~2–3% contributions from each of the other two $M_J = \pm½$ levels [35].

In tetragonal symmetry, the orbital singlet is lowest in energy (B_{2g} ground term), as depicted in Fig. 4b. Because this distortion is along one of the principal (Cartesian) axes of the octahedron, the definition of the orbitals themselves is unchanged from their octahedral counterparts. While the dπ orbital definitions differ between the trigonal and tetragonal examples, electronically, the consequences are simply inverted, within similar limits. It is important to remember that when the distortion is axial, the low-lying ^4E will remain split by spin-orbit coupling, in a fashion similar to the trigonal example above. This affords one new $M_J = \pm^3/_2$ and two new $M_J = \pm^1/_2$ levels that will mix similarly into the ground doublet, which is otherwise determined by the electron spin sub-level within the $^4A_2$ ($M_S = \pm^1/_2$ or $\pm^3/_2$). In the limit that the

$^4A_2$ term is electronically isolated from the ^4E, the ground level is a simple spin quartet. This spin quartet will split, in the absence of a magnetic field, into two doublets ($M_S = \pm^1/_2$ or $\pm^3/_2$) through second-order spin-orbit coupling, according to the zero-field Hamiltonian in Eq. (1) [30]. In the axial limit ($E = 0$), the energy separation corresponds to $2D$ (by convention, a value of $D < 0$ corresponds to $M_S = \pm^3/_2$ lowest in energy, while a value of $D > 0$ corresponds to $M_S = \pm^1/_2$ lowest in energy). Any lower symmetry distortion, in either the trigonally or tetragonally distorted octahedron, will lift the degeneracy of the ^4E level and quench the remaining orbital angular momentum.

$$\hat{H} = D\left[\hat{S}_z^2 - \frac{1}{3}E\left(\hat{S}_x - \hat{S}_y\right)\right] \tag{1}$$

In contrast to the complexities of pseudo-octahedral coordination, a purely tetrahedral ligand field (left side of Fig. 2c) produces a $^4A_2$ electronic ground state. Consequently, pseudotetrahedral complexes are readily described as a zero-field split, spin quartet ground state. However, it is important to remember that the $^4T_2$ excited levels will remain under the influence of spin-orbit coupling in first order (this is what gives rise to the multiplet structure often observed on the visible transition for tetrahedral Co(II) complexes) [29]. Distortions away from tetrahedral symmetry, both toward trigonal symmetry (e.g., elongation along one bond) and toward tetragonal symmetry (elongation along the axis that bisects one L-Co-L angle), have been predicted to favor an $M_S = \pm^3/_2$ ground doublet, while similar compressions are predicted to favor an $M_S = \pm^1/_2$ ground doublet [36].

Translation to EPR Properties and ZFS in High-Spin Co(II)

In each case discussed above, an M_S (M_J) = $\pm^1/_2$ level is available that should be easily accessible by EPR, and aside from the trigonally-distorted octahedron, the zero-field splitting (ZFS) Hamiltonian of Eq. (1) is expected to be sufficient to describe a system of low symmetry. In 1978, Pilbrow formalized the effective *g*-value formalism that is still used today [37]. By expanding the eigenvalue expressions to third order in the ratio of $D/h\nu$, analytical expressions for the dependence of the effective *g*-values on the axial (D) and rhombic (E) ZFS energies were derived, leading to the well-known relations in Eq. (2), where $D' = |\Delta| = [D^2 + 3E^2]^{1/2}$ [37]. This formalism was experimentally extended by Gatteschi and co-workers in a series of seminal works that appeared between 1977 and 1982 [36, 38–46]. In an angular overlap procedure, they used known crystal structures and parameters derived from fitting the optical spectra to fully diagonalize the 40 × 40 matrix derived from the ^4F and ^4P free ion terms, calculating the experimental g^{eff} values directly. However, they then returned to the perturbation-based Eq. (2) to relate their effective *g*-values from the $S' = ^1/_2$ Hamiltonian to the real *g*-values and zero-field splittings of the $S = ^3/_2$ Hamiltonian.

In the axial limit, this model predicts $g_{\parallel}^{eff} = g_{\parallel}$ and $g_{\perp}^{eff} = 2g_{\perp}$ for the $\tilde{M}_S = \pm^1/_2$ doublet and $g_{\parallel}^{eff} = 3g_{\parallel}$ and $g_{\perp}^{eff} = 0$ for the $\tilde{M}_S = \pm^3/_2$ doublet [37]. For isotropic $g = g_e$, this gives rise to the familiar $g_{\perp}^{eff} = 4$ and $g_{\parallel}^{eff} = 2$ signal seen for other $S = ^3/_2$ systems with a well-documented $\tilde{M}_S = \pm^1/_2$ ground level, such as the ferrous-nitrosyl ($[FeNO]^7$) complex of non-heme Fe(II) enzymes [47], and octahedral Cr(III). As rhombicity is introduced ($^E/_D > 0$), the perpendicular g-values will split proportionally, while $g_{\parallel}$ shifts to lower values for both doublets. As noted in the previous review in this series, [5] if one assumes axial symmetry in the real g-values (always reasonable for a first row transition metal), and further assumes that only one doublet is thermally accessible, the result is three observable g-values for three parameters ($g_{\perp}$, $g_{\parallel}$, and $^E/_D$) in Eq. (2). The biggest advantage of Eq. (2) is that no prior knowledge of the spin system is necessary, as the axial ZFS (D) only enters the equations as part of the ratio $^E/_D$.

$$\tilde{M}_S = \pm 1/2 \qquad \tilde{M}_S = \pm 3/2$$

$$g_x^{eff} = g_x\left[1 + \frac{1-3\left(\frac{E}{D}\right)}{\sqrt{1+3\left(\frac{E}{D}\right)^2}}\right] \qquad g_x^{eff} = g_x\left[1 - \frac{1-3\left(\frac{E}{D}\right)}{\sqrt{1+3\left(\frac{E}{D}\right)^2}}\right] \tag{2A}$$

$$g_y^{eff} = g_y\left[1 + \frac{1+3\left(\frac{E}{D}\right)}{\sqrt{1+3\left(\frac{E}{D}\right)^2}}\right] \qquad g_y^{eff} = g_y\left[1 - \frac{1+3\left(\frac{E}{D}\right)}{\sqrt{1+3\left(\frac{E}{D}\right)^2}}\right] \tag{2B}$$

$$g_z^{eff} = g_z\left[1 - \frac{2}{\sqrt{1+3\left(\frac{E}{D}\right)^2}}\right] \qquad g_z^{eff} = g_z\left[1 + \frac{2}{\sqrt{1+3\left(\frac{E}{D}\right)^2}}\right] \tag{2C}$$

It is, however, important to remember the limitations of Eq. (2) as we move forward. The first is that the labels $M_S = \pm^3/_2$ and $M_S = \pm^1/_2$ are only appropriate when the system is perfectly axial ($^E/_D = 0$). Any non-zero value for $^E/_D$ has the effect of mixing the spin sub-levels ($\tilde{M}_S$), proportionally, up to the rhombic limit ($^E/_D = 0.33$). At the rhombic limit, the levels are completely quantum mechanically mixed, and this is the source of the term "rhombic Fe" for a high-spin Fe(III) which shows an isotropic signal at $g^{eff} \sim 4.3$; similarly, there is the potential for a "rhombic Co" signal at $g^{eff} \sim 3.3$ [30]. Any increase in the value of $^E/_D$ beyond 0.33 is mathematically equivalent to a re-labeling of the axes, while maintaining $0 \leq {}^E/_D \leq 0.33$.

We wish to point out here that it is not necessary to re-label the axes to simulate an $\tilde{M}_S = \pm^1/_2$ level and its corresponding $\tilde{M}_S = \pm^3/_2$ level, as has been suggested [5, 43]. It is common practice to simulate *hs* Co(II) EPR spectra with an axial g-tensor while setting $g_x = g_{\|}$, in order to account for ^{59}Co hyperfine coupling, only resolved at g_x^{eff}, and only predicted at g_z^{eff}. The observation was described in the original work as "… the somewhat unexpected feature of a quasi-axial *A* tensor, but with the unique $A_{\|}$ value associated with one of the $g_{\perp}$ components" [43]. While theory was developed to account for the observation, it is important to remember that permuting the *g*-values is equivalent to rotating one doublet's *g*-tensor relative to the other. In other words, this assumes the spin sub-levels' axes of quantization are offset by 90° (the same effect can be accomplished in a simulation with $g_z = g_{\|}$, by including a 90° rotation in *D*). While there is NMR evidence to suggest the spin sub-levels are quantized differently in fluid solution [48], there is, to our knowledge, no evidence this should be true in frozen solution. As we will show below, we believe the ^{59}Co hyperfine coupling in question originates in the $\tilde{M}_S = \pm^3/_2$ level, and its appearance is indicative of an $\tilde{M}_S = \pm^3/_2$ ground level.

The most important thing to remember about Eq. (2) is that it was derived in the limit of $D \gg h\nu$ (in simulating *hs* Co(II) EPR, it is typical to set *D* to an arbitrarily large value of 50 cm^{-1} or more). This practice perhaps gained acceptance in light of a now infamous set of power saturation studies that attempted to correlate X-band temperature- and power-dependent EPR of a set of known model compounds and reasonably well-characterized Co(II)-substituted metalloenzymes to the ZFS [49, 50]. Based on Eq. (3), which assumes a single-phonon, Orbach relaxation mechanism [51], the method was shown to work well for a ferric heme, for which independent measurements were available [50]. The Co(II) method fails, we believe, for assuming that the electron relaxation at low temperature is well-described as a single-phonon process, which necessitates the assumption that the ground level is $\tilde{M}_S = \pm^1/_2$. As we show below, we believe that the ground sub-level in many of these cases is, in fact, $\tilde{M}_S = \pm^3/_2$, which would require multi-phonon (Raman) processes, or multiple, simultaneous Orbach processes, to affect electron relaxation at low temperature. For those that show an $\tilde{M}_S = \pm^1/_2$ ground level, as discussed above, nearly every geometry leaves one or more excited levels close enough in energy to mix directly into the ground level wavefunctions, opening pathways for direct relaxation.

$$P_{1/2} = \frac{C}{T_1} = Ce^{-\Delta/kT} \quad (3)$$

There is general agreement that variable-temperature MCD and high-field/frequency EPR (HF/F-EPR) are the most reliable methods to determine ZFS parameters experimentally for *hs* Co(II), but there are only a few relevant studies available [4, 10, 14, 15]. A discussion of the relative merits of the two methods is well outside the scope of this chapter, and we have limited this review to mononuclear complexes [52, 53]. However, HF/F-EPR has the advantage of being able to determine

Table 1 Summary of available model-derived ZFS parameters from MCD and HF/F-EPR

Complex	Donors	HF/F-EPR				MCD	Refs.
		g	D	$^E/_D$	$\lvert\Delta\rvert^a$	$\lvert\Delta\rvert^a$	
Four-coordinate							
1. CoL_2Cl_2 (L = Me_2nitroimidazole)	N_2Cl_2	[2.37, 2.09, 2.16]	+11.4	0.21	24.0		[15]
(1) (L = imidazole)	N_2Cl_2	[2.45]	+9.2	0.11	18.7	5.3	[4, 15]
(2) (L = benzimidazole)	N_2Cl_2	[2.22, 2.24, 2.24]	±3.3	0.28	7.4		[15]
(3) (L = cytosine)	N_2Cl_2	[2.23, 2.22, 2.25]	−4.3	0.05	8.7		[15]
(4) (L = quinoline)	N_2Cl_2	[2.21, 2.20, 2.19]	−5.9	0.26	13.0		[15]
(5) (L = aminopyrimidine)	N_2Cl_2	[2.25, 2.23, 2.20]	−8.0	0.28	17.7		[15]
(6) (L = PPh_3)	P_2Cl_2	[2.17, 2.17, 2.24]	−14.8	0.08	30	30[b]	[10]
(7) $Tp^{t\text{-Bu},H}Co(NCS)$	N_4	[2.27, 2.28, 2.21]	+2.4	0.06	4.8	3.7	[4, 14]
(8) $Tp^{t\text{-Bu},Me}Co(NCS)$	N_4	[2.16, 2.18, 2.23]	+2.7	0.06	5.4	3.1	[4, 14]
(9) $Tp^{t\text{-Bu},Tn}Co(NCS)$	N_4	[2.24, 2.27, 2.23]	+3.3	0.15	6.9	3.7	[4, 14]
(10) $Tp^{t\text{-Bu},H}Co(NCO)$	N_4	[2.29, 2.43, 2.16]	+6.0	0.03	12.0	3.0	[4, 14]
(11) $Tp^{t\text{-Bu},Me}Co(NCO)$	N_4	[2.16, 2.27, 2.22]	+5.4	0.08	10.8	3.0	[4, 14]
(12) $Tp^{t\text{-Bu},H}Co(N_3)$	N_4	[2.48, 2.02, 2.31]	+7.5	0.21	15.9		[14]
(13) $Tp^{t\text{-Bu},Me}Co(N_3)$	N_4	[2.70, 2.46, 2.18]	+6.3	0.08	12.8	4.2	[4, 14]
(14) $Tp^{t\text{-Bu},H}CoCl$	N_3Cl	[2.34, 2.32, 2.11]	+10.9	0.02	21.8		[14]
(15) $Tp^{t\text{-Bu},Me}CoCl$	N_3Cl	[2.35, 2.35, 2.24]	+11.5	0.01	23.0	4.8	[4, 14]
(16) $Tp^{t\text{-Bu},Tn}CoCl$	N_3Cl	[2.36, 2.20, 2.25]	+12.7	0.06	25.6		[14]
(17) $[(2Me\text{-}Im)_4Co](ClO_4)_2$	N_4					5.1	[4]
(18) $[(2Me\text{-}Im)_4Co](BF_4)_2$	N_4					4.1	[4]
(19) $Co(B(3\text{-isopropylpyrazole})_4)_2$	N_4					2.3	[4]
(20) $Co(bisimidazolylbiphenyl)_2$	N_4					5.6	[4]
(21) $Co(thiourea)_2Cl_2$	S_2Cl_2					3.8	[4]
(22) $Co[SC(CH_3)_2CH_2NH_2]_2$	S_2N_2					21	[4]
Five-coordinate							
(23) $Co(Et_4dien)Cl_2$	N_3Cl_2					98	[4]
(24) $[Co(Me_6tren)Cl]Cl$	N_4Cl					5.4	[4]
(25) $[Co(Me_6tren)NCS]NCS$	N_5					2.7	[4]
(26) $Tp^{t\text{-Bu},}HCo(NO_3)$	N_3O_2					6.9	[4]
(27) $Tp^{t\text{-Bu},}TnCo(BBN(pz)_2)$	N_5					16	[4]
(28) $Tp^{t\text{-Bu},}MeCo(NCS)(THF)$	N_4O					14	[4]

[a] $\lvert\Delta\rvert = \lvert 2D[1 + 3(^E/_D)^2]^{1/2}\rvert$
[b] From VT/VH-MCD

both the sign and magnitude of the ZFS parameters, while VT-MCD can only determine their magnitude. The available data are summarized below in Table 1.

As is clear from Table 1, where there is overlap between independent measurements, there is generally poor agreement between the two techniques, aside from entries 8–10. Of the 29 complexes in Table 1 (23 four-coordinate and 6 five-coordinate), 13 include at least one chloride ligand. This includes 11 of the 23 four-coordinate complexes, while 7 more (entries 8–14, N_4 donor set) replace the halide with a pseudohalide (NCS^-, NCO^- N_3^-), leaving just five entries [18–21, 23] with donor sets that approximate those likely to be encountered in a Co(II)-substituted metalloenzyme (four tetra-azole complexes and a bisthioether/bisamine). No five- or six-coordinate complexes have been studied in detail by HF/F-EPR (spectra of a six-coordinate [(R-pyridine)$_4$Co(NCS)$_2$] complex incorporated into a MOF, have recently been reported, but a detailed analysis was not included) [54]. Of the 6 five-coordinate complexes that have been studied by MCD (Table 1), one incorporates two chloride ligands, and three others include chloride, thiocyanide and a solvent of crystallization as ancillary ligands. The remaining two complexes [27, 28] are the best approximations of biological metal sites.

The four-coordinate bisthioether/bisamine complex (#23) provides the only VT-MCD-derived value of $|\Delta|$ outside the fairly tight range of $2.3 < |\Delta| < 5.6\ \text{cm}^{-1}$ that defines the remainder (except the Co(PPh$_3$)$_2$Cl$_2$ complex, which was part of a different study). Meanwhile, aside from the dichloro/dien complex ($|\Delta| = 98\ \text{cm}^{-1}$), all of the five-coordinate complexes studied by MCD yield $|\Delta| \leq 16\ \text{cm}^{-1}$, while the two most relevant complexes [27, 28] differ by a factor of two. In contrast, the HF/F-EPR data, mostly acquired between 120 (4 cm^{-1}) and 700 GHz (23.3 cm^{-1}), indicate widely varying values, $7.4 < |\Delta| < 30\ \text{cm}^{-1}$ for tetragonal four-coordinate complexes [1–7] and $4.8 < |\Delta| < 25.6\ \text{cm}^{-1}$ for trigonal four-coordinate complexes [8–17]. If the HF/F-EPR values are closer to the truth, these measurements are far outside either limit ($\Delta \gg h\nu$, as expected for Eq. (2), or $\Delta \ll h\nu$, the goal of HF/F-EPR), being closer to $\Delta \sim h\nu$. This may explain some of the variations in Table 1, within sets of otherwise highly similar complexes (compare entries 1–6 and 8–17). For example, the azole complexes [1–6] all contain a tetragonal structure, and all yielded approximately isotropic *g*-tensors (except the nitroimidazole complex). What is striking is within complexes 2–6 in Table 1 is that *D* ranges from +9.2 to $-8.0\ \text{cm}^{-1}$, while $^E/_D$ ranges from 0.05 to 0.28 (3 of 5 were very near the rhombic limit ($^E/_D > 0.26$). Similarly, within the Tp complexes 8–17, the g-anisotropy shows no obvious trends, either within sets that bear the same Tp ligand or those that bear the same ancillary ligand. While $|\Delta|$ appears to increase with the π-donor ability of the ancillary ligand, from NCS$^-$ to NCO$^-$ to N$_3^-$ to Cl$^-$, there is no obvious reason why it should lead $|\Delta|$ to increase by a factor of 4 or 5.

In summary, we believe the question of the ground spin state, and the influence of spin-orbit coupling and ZFS, remains an open one. Given that $k\text{T} \sim 0.69\ \text{cm}^{-1}/\text{K}$, a value of $|\Delta|$ in the MCD-derived range for four-coordination given above *requires* substantial population of both doublets, even at the lowest temperatures. For example, even at 2 K ($k\text{T} \sim 1.4\ \text{cm}^{-1}$), a complex with $|\Delta| = 2.3\ \text{cm}^{-1}$ will have an 84/16 population distribution (for $|\Delta| = 5.6\ \text{cm}^{-1}$, the distribution is 98/2 at 2 K). At temperatures more commonly employed in EPR, this can only shift to favor the upper doublet, as shown graphically in Fig. 5. Taking the extreme example of $|\Delta| = 30\ \text{cm}^{-1}$ from the HF/F-EPR measurements [10], at 10 K, there is still ~1% population in the upper doublet, so it is important to remember that as the value of $|\Delta|$ decreases, the ability to thermally isolate one doublet from the other also decreases. The result of the above is spin functions that are not well isolated, and at least partially mixed, containing non-negligible quantum mechanical contributions from low-lying electronic excited states, with significant population of both spin doublets likely in

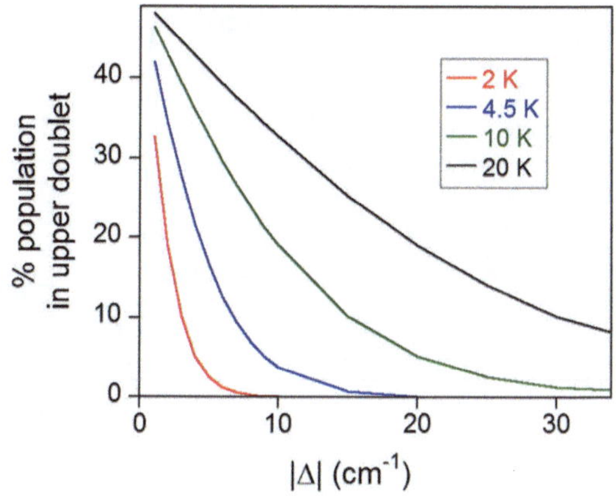

Fig. 5 Percent population residing in the upper spin doublet, as a function of the doublet separation, |Δ|, at several temperatures

many cases. What follows are several example studies from our laboratory in support of the narrative above, and how it affects our ability to probe these systems using multiple resonance techniques.

Examples

Four- and Five-Coordinate Pyrazolylborates

Recently, we reported optical and EPR spectroscopy of a closely related set of *bis*bispyrazolylborate complexes [55], first reported in 1967 [56], and loosely related to entries 8–17 in Table 1. Their structures are described in Fig. 6a. Solution optical and powder EPR studies of the parent complex, Bp_2Co, the $(Bp^{3,5Me})_2Co$ analog, and the boron-substituted $(Ph_2Bp)_2Co$ complex were reported in 1973 [57]. Bp_2Co was also examined as a single-crystal, showing axial, but unusually anisotropic spectra with $g_{\perp}^{eff}$ ~ 1.0 and $g_{\parallel}^{eff}$ ~ 6.9, assigned to an $\tilde{M}_S = \pm^3/_2$ ground state, and consistent with an axially elongated intrinsic g-tensor ($g_{\parallel}$ (~2.3) > $g_{\perp}$). Surprisingly, as illustrated in Fig. 6b [55], the frozen solution spectra are dominated by a new signal with the shape of a more classic $S = {}^3/_2$ paramagnet with an $\tilde{M}_S = \pm^1/_2$ ground state, with $g_{\perp}^{eff}$ = 4.7 and $g_{\parallel}^{eff}$ = 2.0 (a compressed intrinsic g-tensor, $g_{\perp}$ (~2.3) > $g_{\parallel}$(2.0)), consistent with $^E/_D$ ~ 0.005. Closer inspection shows that the signals originally reported in the crystal are also present, at markedly lower intensity. Power saturation shows that the $\tilde{M}_S = \pm^3/_2$ signal disappears very quickly with increasing temperature, while the $\tilde{M}_S = \pm^1/_2$ signal maximizes in intensity at ~6 K, clearly showing the $\tilde{M}_S = \pm^3/_2$ level is lowest in energy; similar studies suggest this is true for the entire series. This example illustrates three key points. The first is that both doublets are populated at 4.5 K. The second is that the study of powders and, in this case, single-crystals, can provide a means for quenching angular momenta that are unaffected in frozen solution, potentially masking resonances. We note here that, at present, HF/F-EPR cannot be measured in frozen solution. Although the

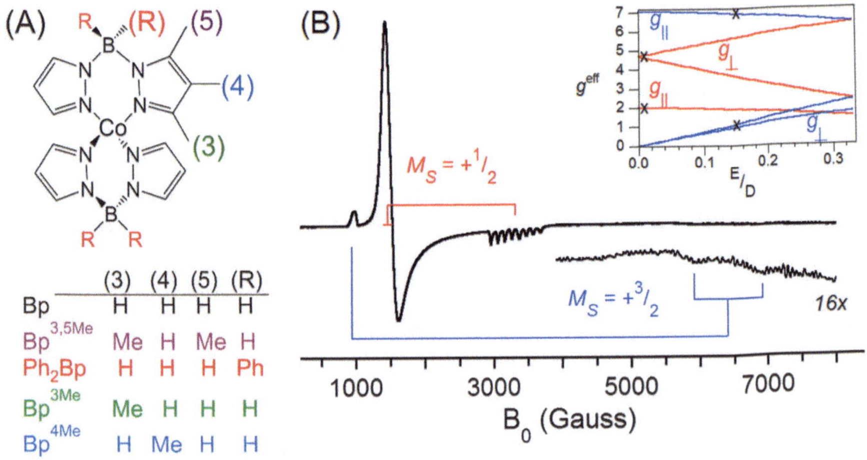

Fig. 6 (**a**) Structure of Bp_2Co. The substitution nomenclature is as indicated in the table below. (**b**) X-band EPR of Bp_2Co in toluene solution at 4.5 K. The *brackets* indicate the assignment of resonances [55]. *Inset*: Rhombogram, based on intrinsic *g*-values given in text and Eq. (2); observed resonance positions are marked with an X

authors of these important studies took great care to account for it, nearly every complex in Table 1 torqued (partially aligned) in the large static fields used.

The third, and perhaps most important point to be made is that Eq. (2) cannot fully explain the *g*-values observed. That is, based on $^{E}/_{D}$ and the *g*-values from the $\tilde{M}_S = \pm^1/_2$ signal, and allowing for the interchange of intrinsic *g*-values, we should expect the $\tilde{M}_S = \pm^3/_2$ signal to only show $g_{\parallel}$ at a position slightly higher in g than observed (~7.1), with $g_{\perp}$ formally 0. Examination of the rhombogram (inset to Fig. 6b) shows that the observed g-values ($g_{x,y,z}(^3/_2)$ = [0.95, 1.13, 6.92]) are matched nicely, but at an $^{E}/_{D}$ of 0.15. In his original report, Pilbrow showed that $g_{\perp}^{eff}$ must be corrected, in the limit of axial symmetry, and that these corrections will become more important in the limit $D \sim h\nu$ [37]. Although we are unable to apply them in the absence of an independent estimate of *D*, based on the values in Table 1, we must anticipate that any correction will be small. For the other complexes in the series where all six transitions were observed, similar arguments can be made [55].

Comparing the data across the series of complexes supports much of the above, and offers some potential insight into Co(II)-substituted metalloprotein spectra, such as those in Fig. 1. The first observation speaks to the appearance of the low-field end of the spectra, where we believe the presence or absence of ^{59}Co hyperfine coupling is dictated by the energy matching of the low-field resonances (g_1) from each doublet. The low-field region of the spectra are presented in Fig. 7a. As the two low-field resonances (g_1 from each doublet) are well separated in Bp_2Co, the ^{59}Co-bearing $\tilde{M}_S = \pm^3/_2$ derived signal is barely observed. In fact, in metalloproteins, a similar signal is often seen as a shoulder to low-field, and ascribed to a low-lying excited state [5]. Here we suggest it is coming from the ground state.

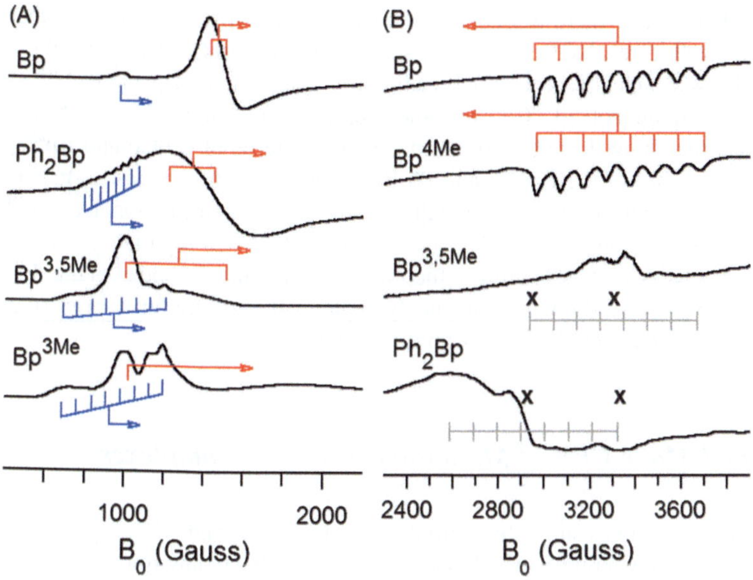

Fig. 7 X-band frozen solution EPR of substituted $(RBp^x)_2Co$ complexes in toluene solution at 4.5 K [55]. (**a**) Expansion of the low-field region. The brackets are meant to guide the reader, marking the assignment of features to $\tilde{M}_S = \pm^1/_2$ or $\tilde{M}_S = \pm^3/_2$. (**b**) Expansion of the $g \sim 2$ region. The gray brackets are the same size as those above for the Bp_2Co ^{59}Co hyperfine pattern, placed arbitrarily over the observed signals. Positions assigned as potential g-values in power spectra are marked with an X [57]

As the two low-field transitions converge, they appear to couple, through an as yet unclear mechanism. We favor this description (arising from a coupling, or an intensity stealing mechanism) based on the similarity of the complexes. Thus, in the Bp^{3Me} and $Bp^{3,5Me}$ derivatives, the $\tilde{M}_S = \pm^1/_2$ derived signal is unusually intense, and at the same time, the $\tilde{M}_S = \pm^3/_2$ derived signal is observed. Based on our inability to detect any of the other transitions reliably for the Bp^{3Me} analog, we propose that the relative intensity of these two contributions may be indicative of the rhombicity, apparent in the $\tilde{M}_S = \pm^1/_2$ component. That is, as g_2 and g_3 of the $\tilde{M}_S = \pm^1/_2$ component are relatively close in the $Bp^{3,5Me}$ derivative, the $\tilde{M}_S = \pm^1/_2$ component at g_1 is more intense. Conversely, as g_2 and g_3 of the $\tilde{M}_S = \pm^1/_2$ component for Bp^{3Me} appear to be much further apart, the $\tilde{M}_S = \pm^1/_2$ and $\tilde{M}_S = \pm^3/_2$ components contribute more equal intensity. The Ph_2Bp derivative would then represent the in-between case, where the two g_1 transitions are just close enough in energy for the $\tilde{M}_S = \pm^3/_2$ derived transition to be observed, with a line shape that is determined by the sum of the two.

Secondly, as suggested by Fig. 7b, we believe this may be more general, applying to many pseudotetrahedral and five-coordinate Co(II) EPR spectra. In most metalloprotein spectra, the g_3 (high-field) transition assigned to the $\tilde{M}_S = \pm^1/_2$ level is very broad and poorly defined. While it has been seen in other model complexes, to our knowledge there are no examples of Co(II)-substituted proteins that show ^{59}Co hyperfine resolved near g ~ 2, such as that of Bp_2Co. However, the data in Fig. 7b make a clear case for the breadth of g_3 (g_z) in the Ph_2Bp and $Bp^{3,5Me}$ systems arising from unresolved ^{59}Co hyperfine coupling. To illustrate, in Fig. 7b we placed brackets matching those that frame the 8-line patterns of Bp_2Co and $Bp^{4Me}{}_2Co$ on the same region of the $Bp^{3,5Me}$ and Ph_2Bp spectra, and in both cases, $A(^{59}Co)$ from the parent complex easily spans the discernible signals in the region near g = 2. Note that the brackets also span what was interpreted as two separate g-values in the original powder studies, leading to their mis-assignment as g_x and g_y of the $\tilde{M}_S = \pm^3/_2$ level. In Fig. 8, we show that this nicely explains the spectra for a series of crystallographically characterized five-coordinate complexes.

Parallel Mode EPR of Mononuclear Co(II) Complexes

The above examples show that both the and $\tilde{M}_S = \pm^1/_2$ and $\tilde{M}_S = \pm^3/_2$ doublets are populated at low temperature. However, as noted above, any non-zero value of $^E/_D$ will lead to proportionate mixing of the $\tilde{M}_S = \pm^1/_2$ and $\tilde{M}_S = \pm^3/_2$ doublets. Proof of such mixing is easily obtained by examining the EPR response with $B_1 \parallel B_0$ (parallel mode) [58]. Unlike conventional EPR ($B_1 \perp B_0$), the selection rule in parallel mode EPR is formally $\Delta M_S = 0$ [59]. Affecting a net absorption of radiation without changing the total spin requires off-diagonal matrix elements that directly connect two or more levels. In the present case, where $D > h\nu$, similar logic can be applied to the doublet splittings in *hs* Co(II) complexes.

Three examples are shown in Fig. 9 [58]. The first (Fig. 9a) is the $Tp^{Ph,Me}Co$(1,2-hopo) complex from Fig. 8. The brackets from Fig. 8 have been retained, and should make it readily apparent that parallel-mode detection accentuates different features.

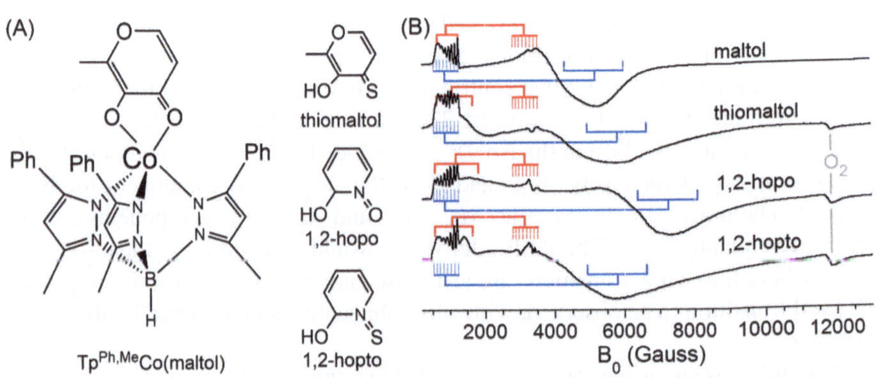

Fig. 8 X-band frozen solution EPR of substituted five-coordinate $Tp^{Ph,Me}CoL$ complexes, where L is a bidentate (O, O) or (O, S) donor. (**a**) Structures. (**b**) X-band EPR in toluene/dichloromethane solution at 3.7 K. The *brackets* indicate proposed assignment of resonances to $\tilde{M}_S = \pm^1/_2$ or $\tilde{M}_S = \pm^3/_2$. The signal at ~11.5 kG derives from dissolved O_2 (g ~ 0.6)

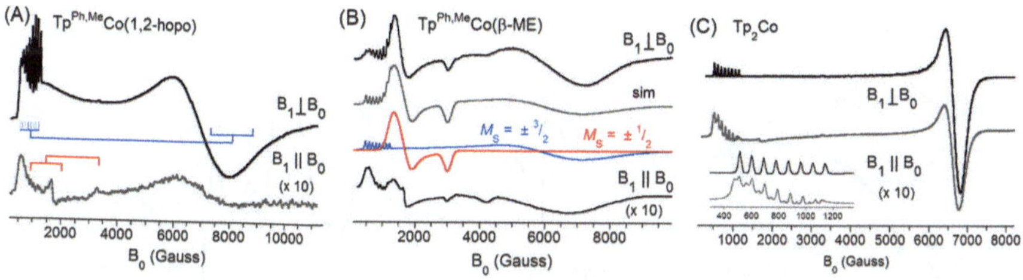

Fig. 9 Perpendicular and parallel mode EPR of (**a**) five-coordinate $Tp^{Ph,Me}Co(1,2\text{-hopo})$, (**b**) five-coordinate $Tp^{Ph,Me}Co(\beta\text{-mercaptoethanol})$ and (**c**) six-coordinate Tp_2Co [58]. Inset: Expansion of the low-field region, showing the presence of ^{59}Co nuclear transitions. The *brackets* indicate proposed assignment of resonances to $\tilde{M}_S = \pm{}^1/_2$ or $\tilde{M}_S = \pm{}^3/_2$

In fact, parallel mode makes it much easier to see the signals from the $\tilde{M}_S = \pm{}^1/_2$ doublet excited state. The most striking example is shown in Fig. 9b, $Tp^{Ph,Me}Co(\beta\text{-mercaptoethanol})$ (BME). Included in Fig. 9b is a total simulation, generated by adding equal components of $\tilde{M}_S = \pm{}^1/_2$ and $\tilde{M}_S = \pm{}^3/_2$. The simulations (red and blue traces in Fig. 9b) were generated using the same g-tensor ($g_{x,y,z} = [2.25, 2.25, 2.63]$), and reversing the sign of both D and E, *without permuting any g-values* (as the parameter $^E/_D$ is only defined as a positive quantity, it is important to remember that failure to change the sign of *both D* and E will lead to a requirement that axes be permuted). This is the only complex for which we have had such straightforward success. Again, the parallel mode signal from this complex makes it easier to visualize the full set of transitions.

The third example, in Fig. 9c, shows what happens when there is not significant sub-level mixing. No new signals appear, and the signal that is observed is by far the weakest we have seen. As discussed below, this is the most electronically isolated ground state we have looked at, and the signal intensity in parallel mode is less than 3% that of the perpendicular signal (compared to >11% for the two five-coordinate examples in Fig. 9). Although it is a loose correlation, the 3% intensity matches well with the amount of excited state mixing expected in this complex [35]. At high power, the largest signals observed are for five-coordination, while at low power, signal intensities track almost linearly, four-coordinate > five-coordinate > six-coordinate [58]. We have reported similar signals for a number of other Co(II) complexes, with a wide range of structures and donor sets [58, 60, 61]. In fact, thus far, we have yet to find a *hs* Co(II) complex that does not give a response in parallel mode, indicating the prevalence of spin sub-level mixing in these systems.

ENDOR of Six-Coordinate Trigonal Co(II)

The final example we wish to discuss in some detail is the *bis*-trispyrazolylborates of Co(II). Although a trigonally distorted octahedron is rarely encountered in biochemistry, this system remains the only set of Co(II) complexes thoroughly studied

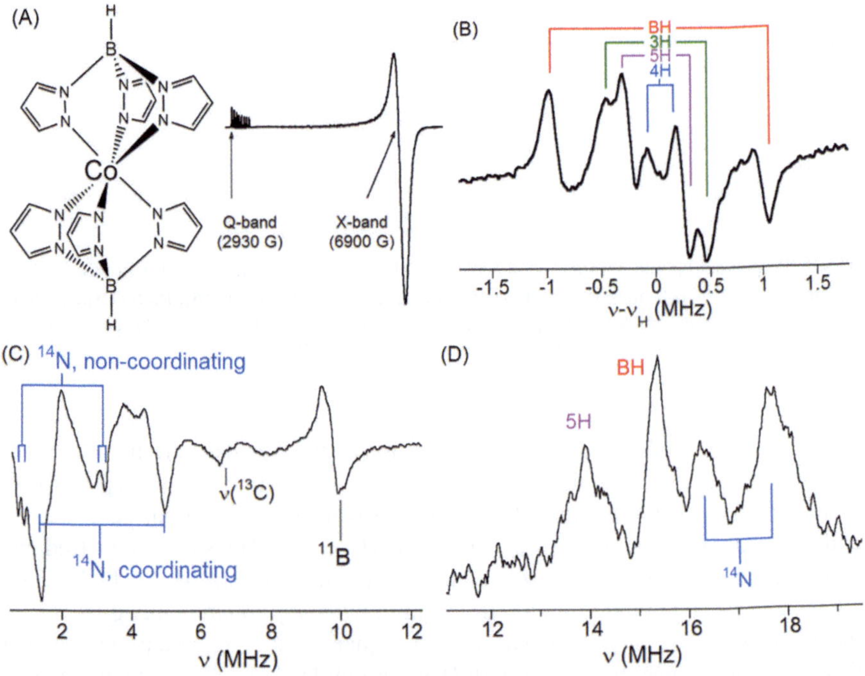

Fig. 10 CW ENDOR of Tp_2Co. (**a**) Structure and X-band EPR spectrum, showing field positions of the ENDOR spectra that follow. (**b**) X-band ^1H ENDOR at g ~ 1.0 (ν_H ~ 30 MHz). *Colors* follow the same scheme as in Fig. 6a [8]. (**c**) X-band, low-frequency region at g ~ 1.0 [8]. (**d**) Q-band ENDOR at g ~ 8.5 [9]

using ENDOR spectroscopy [8, 9]. Somewhat surprisingly, to date there remains only two CW ENDOR studies [8, 9], one pulsed-EPR/ESEEM study [62] and one pulsed-ENDOR study [7] in the literature. The trigonal elongation in the present complexes ($r_{N\text{-}N}$ interligand ~10% longer than $r_{N\text{-}N}$ intraligand, Fig. 10a) was first examined by Jesson [34], who gave a careful description of the multiplet structure (see Fig. 3a) through a simultaneous analysis of the room temperature optical and low-temperature EPR spectra. The four lower doublets are separated by ~220 cm^{-1} from one level to the next, and the $^4A_2$ lies ~1900 cm^{-1} above the $^4E(\delta)$. At 220 cm^{-1}, the ground $M_J = \pm{}^1/_2$ doublet is far removed from the nearest excited state, and is ~97% $|\mp 1, \pm\frac{3}{2}>$ [8, 35]. The presence of a full unit of orbital angular momentum in the ground level leads to wildly anisotropic g-values, $g_{\parallel}^{eff}$ ~ 8.5 and $g_{\perp}^{eff}$ ~ 1.0, as shown above in Fig. 9c and in Fig. 10a.

We have reported CW ^{11}B, ^{14}N and ^1H ENDOR of this complex at the two extremes of the EPR signal, as shown in Fig. 10b, c, d [8, 9]. X-band CW ENDOR at the high-field ($g_{\perp}^{eff}$ ~ 1.0) end of the spectrum showed ^1H signals well-separated from ^{14}N. This EPR signal would appear at ~2.8 kG, and is consequently inaccessible, at Q-band. Perhaps not surprisingly, the EPR intensity at $g_{\parallel}^{eff}$ ~ 8.5 is too weak to allow detection of CW ENDOR at X-band, while it was possible at Q-band. At $g_{\|}^{eff}$ ~ 8.5 (~3.2 kG), the Larmor frequency places ^{1}H signals directly in the same range as coordinated ^{14}N. Analysis of these data showed that the intrinsic Co-^{14}N hyperfine couplings are in-line with the only report of Co-^{14}N(His) couplings in a metalloprotein [7], while the unusually large *g*-anisotropy leads to anisotropic Fermi couplings [9]. These studies also demonstrated that the B—H bond in this system is electron rich (~30% greater than the three B—N bonds), helping explain how this ligand can share a single negative charge equally among three nitrogens. We have further used the ENDOR to verify that the NMR chemical shifts can be separated into isotropic and dipolar components by comparing simple structural modifications [8].

Prospects for Pulsed EPR

Much of the above discussion concerns itself with the nature of the Co(II) ground spin state. As noted in the previous treatment in this series [5], conditions can nearly always be found where the $\tilde{M}_S = \pm^1/_2$ signal dominates the spectrum. Thus, if an $\tilde{M}_S = \pm^3/_2$ ground state is encountered, raising the temperature puts population in the $\tilde{M}_S = \pm^1/_2$ energy level, where the transition probabilities are orders of magnitude higher, which will inevitably simplify the observed signal. However, the nature of the ground state is a critical question for the use of multiple resonance techniques and pulsed excitation, which typically cannot be performed at elevated temperatures. The above example, where the magnetic ground state is fairly pure and well isolated energetically, provides a good test case.

At 4.5 K, and g ~ 1.0 at X-band, the spin-echo-detected T_1 for this complex is a short 33 μs, while T_2 is just 1.7 μs, for both the ^{14}N and ^{15}N isotopologs. Figure 11a, shows *uncorrected* 2-pulse ESEEM time traces (only normalized for presentation),

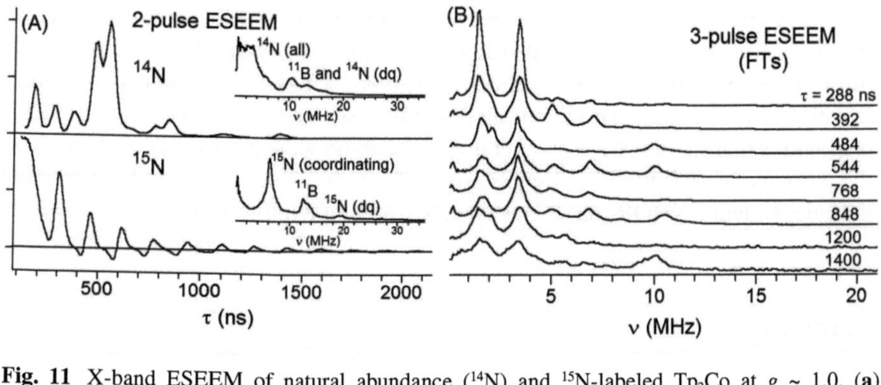

Fig. 11 X-band ESEEM of natural abundance (^{14}N) and ^{15}N-labeled Tp_2Co at g ~ 1.0. (**a**) Uncorrected 2-pulse ESEEM time traces. Insets: Corresponding Fourier transforms. (**b**) Natural abundance Tp_2Co at g ~ 1.0, 3-pulse ESEEM Fourier transforms; τ values as indicated

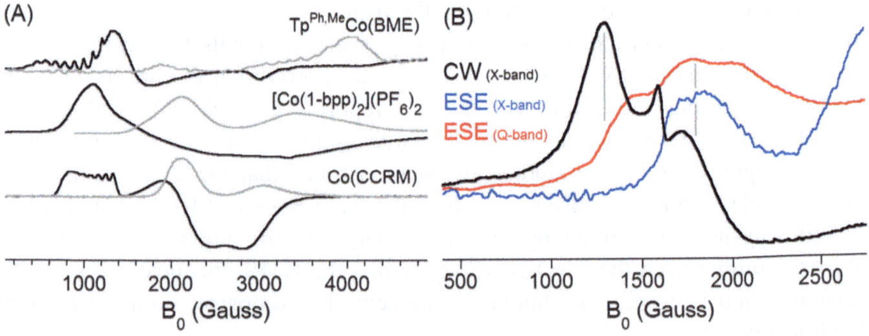

Fig. 12 (**a**) Comparison of X-band CW (*black traces*) and ESE-detected (*gray traces*) of several *hs* Co(II) complexes at 4.5 K. (**b**) Comparison of the low-field region of X-band CW (*black*), X-band ESE-detected (*blue*) and Q-band ESE-detected (*red*) EPR of CoCA, in complex with thiomaltol at 4.5 K

and their corresponding FTs. With six equivalent coordinated ^{14}N, the modulation depth for the natural abundance complex is nearly 100% at regular intervals, resulting in truncation of the FT out to ~4 MHz (see inset), while the ^{15}N complex is well-behaved. However, with careful choice of conditions, high quality 3-pulse ESEEM (Fig. 11b) data can be easily obtained from the natural abundance complex, matching well the ENDOR in Fig. 10c. We note, however, that while we can easily generate simulations that match one, or the other set of spectra (^{14}N or ^{15}N), we have been as yet unable to reconcile the two data sets.

More asymmetric complexes, again perhaps not surprisingly, appear to generally favor the $\tilde{M}_S = \pm^1/_2$ level, as illustrated in Fig. 12, regardless of the ground spin doublet. The six-coordinate bpp and CCRM complexes in Fig. 12a appear to show $\tilde{M}_S = \pm^1/_2$ and $\tilde{M}_S = \pm^3/_2$ ground levels, respectively, and their ESE responses are substantially stronger than the five-coordinate BME complex from above. The BME complex also shows an $\tilde{M}_S = \pm^3/_2$ ground level, while still only showing $\tilde{M}_S = \pm^1/_2$ signals in the ESE-detected spectrum. The same is seen in the thiomaltol complex of CoCA (five-coordinate, Fig. 12b), where the X-band ESE-response is weak. As should be expected, the ESE-response is better at Q-band, and surprisingly information from the $\tilde{M}_S = \pm^3/_2$ level is restored. All of these systems show very fast electron relaxation and very weak ENDOR responses, even when the ESE-EPR response is reasonable. We are as yet unclear as to how the mixing of sub-levels and the presence of multiple EPR signals arising from the same species will complicate field-dependent ENDOR, but this is an active area of exploration.

Summary

The above review of electronic structure, and its translation to EPR-observables, for high-spin Co(II) complexes was intended to aid the general practitioner in unraveling the often complex spectra of these systems. We show how a combination of electronic excited state and spin sub-level mixing, and Boltzmann statistics, lead to difficulty in determining the nature of the magnetic ground state ($\tilde{M}_S = \pm^1/_2$ or $\tilde{M}_S = \pm^3/_2$). While MCD and/or high-field/frequency EPR can do so more definitively, available studies are limited, and the field will benefit from expansion of this database, particularly into frozen solutions. Data are presented that make the case that EPR signals from both doublets are often present at the same time. The relation of the ground state to electron-spin-echo detection is also examined, showing preferentially $\tilde{M}_S = \pm^1/_2$ signals at X-band, while showing signals for both species at Q-band.

Acknowledgements This work was supported by the US National Science Foundation (CHE-0518189, CHE-0964806, CHE-0909985 and CHE-1152755 to D.L.T.).

References

1. Vallee BL, Rupley JA, Coombs TL, Neurath H (1958) The release of zinc from carboxyeptidase and its replacement. J Am Chem Soc 80:4750–4751
2. Maret W, Vallee BL (1993) Cobalt as probe and label of proteins. In: Methods in enzymology, vol 226. Academic, New York, NY, pp 52–71
3. Schenk G, Tierney DL (2014) X-ray absorption spectroscopy of dinuclear hydrolases. Biophys J 107:1263–1272
4. Larrabee JA, Alessi CM, Asiedu ET, Cook JO, Hoerning KR, Klingler LJ, Okin GS, Santee SG, Volkert TL (1997) Magnetic circular dichroism spectroscopy as a probe of geometric and electronic structure of cobalt(II)-substituted proteins: ground-state zero-field splitting as a coordination number indicator. J Am Chem Soc 119:4182–4196
5. Bennett B (2010) EPR of cobalt-substituted zinc enzymes. In: Bological magnetic resonance, vol 29. Springer, New York, NY, pp 345–370
6. Bertini I, Luchinat C, Parigi G (2001) Solution NMR of paramagnetic molecules. Amsterdam, Elsevier
7. Walsby CJ, Krepkiy D, Petering DH, Hoffman BM (2003) Cobalt-substituted zinc finger 3 of transcription factor IIIA: interactions with cognate DNA detected by ^{31}P ENDOR spectroscopy. J Am Chem Soc 125:7502–7503
8. Myers WK, Duesler EN, Tierney DL (2008) Integrated paramagnetic resonance of high-spin Co(II) in axial symmetry: chemical separation of dipolar and contact electron-nuclear couplings. Inorg Chem 47:6701–6710
9. Myers WK, Scholes CP, Tierney DL (2009) Anisotropic fermi couplings due to large unquenched orbital angular momentum: Q-band ^1H, ^{14}N and ^{11}B ENDOR of bis(trispyrazolylborate) cobalt(II). J Am Chem Soc 131:10421–10429
10. Krzystek J, Zvyagin SA, Ozarowski A, Fiedler AT, Brunold TC, Telser J (2004) Definitive spectroscopic determination of zero-field splitting in high-spin cobalt(II). J Am Chem Soc 126:2148–2155
11. Krzystek J, Zvyagin SA, Ozarowski A, Trofimenko S, Telser J (2006) Tunable-frequency high-field electron paramagnetic resonance. J Magn Reson 178:174–183
12. Lawrence J, Beedle CC, Yang E-C, Ma J, Hill S, Hendrickson DN (2007) High frequency electron paramagnetic resonance (HFEPR) study of a high spin Co(II) complex. Polyhedron 26:2299–2303
13. Maganas D, Milikisyants S, Rijnbeek JMA, Sottini S, Levesanos N, Kyritsis P, Groenen EJJ (2010) A multifrequency high-field electron paramagnetic resonance study of $CoIIS_4$ coordination. Inorg Chem 49:595–605

14. Krzystek J, Swenson DC, Zvyagin SA, Smirnov D, Ozarowski A, Telser J (2010) Cobalt(II) "scorpionate" complexes as models for cobalt-substituted zinc enzymes: electronic structure investigation by high-frequency and field electron paramagnetic resonance spectroscopy. J Am Chem Soc 132:5241–5253
15. Idesicova M, Titis J, Krzystek J, Boca R (2013) Zero-field splitting in pseudotetrahedral Co(II) complexes: a magnetic, high-frequency and -field EPR, and computational study. Inorg Chem 52:9409–9417
16. Semproni SP, Milsmann C, Chirik PJ (2014) Four-coordinate cobalt pincer complexes: electronic structure studies and ligand modification by homolytic and heterolytic pathways. J Am Chem Soc 136:9211–9224
17. Chiang L, Allan LEN, Alcantara J, Wang MCP, Storr T, Shaver MP (2014) Tuning ligand electronics and peripheral substitution on cobalt salen complexes: structure and polymerisation activity. Dalton Trans 43:4295–4304
18. Esswein AJ, Surendranath Y, Reecea SY, Nocera DG (2011) Highly active cobalt phosphate and borate based oxygen evolving catalysts operating in neutral and natural waters. Energy Environ Sci 4:499–504
19. Singh A, Spiccia L et al (2013) Coord Chem Rev 257:2607–2622
20. Bloomfield AJ, Sheehan SW, Collom SL, Crabtree RH, Anastas PT (2014) A heterogeneous water oxidation catalyst from dicobalt octacarbonyl and 1,2-bis(diphenylphosphino) ethane. New J Chem 38:1540–1545
21. Ruamps R, Batchelor LJ, Guillot R, Zakhia G, Barra A-L, Wernsdorfer W, Guihery N, Mallah T (2014) Ising-type magnetic anisotropy and single molecule magnet behaviour in mononuclear trigonal bipyramidal Co(II) complexes. Chem Sci 5:3418–3424
22. Li J, Zhang L, Xu G-C, Yu W-X, Jia D-Z (2014) A carbohydrazone based tetranuclear Co(II) complex: self-assembly and magnetic property. Inorg Chem Comm 45:40–43
23. Amjad A, Minguez Espallargas G, Liu J, Clemente-Juan JM, Coronado E, Hill S, del Barco E (2013) Single-crystal EPR spectroscopy of a Co(II) single-chain magnet. Polyhedron 66:218–221
24. Chorazy S, Nakabayashi K, Imoto K, Mlynarski J, Sieklucka B, Ohkoshi S-I (2012) Conjunction of chirality and slow magnetic relaxation in the supramolecular network constructed of crossed cyano-bridged CoII-WV molecular chains. J Am Chem Soc 134:16151–16154
25. Condon EU, Shortley GH (1935) The theory of atomic spectra. Cambridge University Press, London
26. Griffith JS (1961) The theory of transition metal ions. Cambridge University Press, London
27. Ballhausen CJ (1962) Introduction to ligand fields. McGraw-Hill, New York, NY
28. Abragam A, Bleaney B (1970) Electron paramagnetic resonance of transition ions, 2nd edn. Clarendon Press, Oxford
29. Lever ABP (1984) Inorganic electronic spectroscopy. Elsevier, Amsterdam
30. Mabbs FE, Collison D (1992) Electron paramagnetic resonance of d transition metal compounds. Elsevier, Amsterdam
31. Drago RS (1992) Physical methods for chemists. Surfside, Gainesville, FL
32. Shriver DF, Atkins P, Langford CH (1994) Inorganic chemistry, 2nd edn. W.H. Freeman, New York, NY
33. Orgel LE (1955) Spectra of transition-metal complexes. J Chem Phys 23:1004–1014
34. Jesson JP (1966) Optical and paramagnetic resonance spectra of some trigonal Co(II) chelates. J Chem Phys 45:1049–1056
35. Tierney DL (2012) Jahn-Teller dynamics in a series of high-symmetry Co(II) chelates determine paramagnetic relaxation enhancements. J Phys Chem A 116:10959–10972
36. Banci L, Bencini A, Benelli C, Gatteschi D (1980) ESR spectra of low symmetry high spin cobalt(II) complexes. 9. Theoretical considerations on tetrahedral and five coordinate complexes. Nouv J Chim 4:593–598
37. Pilbrow JR (1978) Effective g values for S = $^3/_2$ and S = $^5/_2$. J Magn Reson 31:479–490
38. Bencini A, Gatteschi D (1977) Electron spin resonance spectra of low-symmetry pseudotetrahedral high-spin cobalt(II) complexes. Tetra-n-butylammonium tribromo(quinoline) cobaltate(II). Inorg Chem 16:2141–2142

39. Bencini A, Benelli C, Gatteschi D, Zanchini C (1979) ESR spectra of low-symmetry high-spin cobalt(II) complexes. 3. Square-pyramidal nitratotetrakis(methyldiphenylarsine oxide) cobalt(II) nitrate. Inorg Chem 18:2526–2528
40. Bencini A, Benelli C, Gatteschi D, Zanchini C (1979) ESR spectra of low-symmetry high-spin cobalt(II) complexes. 2. Pseudotetrahedral dichlorobis(triphenylphosphine oxide)cobalt(II). Inorg Chem 18:2137–2140
41. Bencini A, Benelli C, Gatteschi D, Zanchini C (1980) ESR spectra of low symmetry high spin cobalt(II) complexes. 8. Observation of ESR spectra of dichloro-tetrakispyrazole cobalt(II) doped into paramagnetic nickel(II) analog. Inorg Chim Acta 45:L127–L128
42. Bencini A, Benelli C, Gatteschi D, Zanchini C (1980) ESR spectra of low-symmetry high-spin cobalt(II) complexes. 7. Trigonal-bipyramidal pentakis(picoline N-oxide)cobalt(II) perchlorate. Inorg Chem 19:3839–3841
43. Bencini A, Benelli C, Gatteschi D, Zanchini C (1980) ESR spectra of low-symmetry high-spin cobalt(II) complexes. 6. 6-Methylquinoline, pyridine, and water adducts of cobalt(II) acetylacetonate. Inorg Chem 19:3027–3030
44. Bencini A, Benelli C, Gatteschi D, Zanchini C (1980) ESR spectra of low-symmetry high-spin cobalt(II) complexes. Part 5. Pseudotetrahedral tetra-n-butylammonium tribromo(quinoline) cobaltate(II) and the calculation of the A2 tensor. J Mol Struct 60:401–404
45. Bencini A, Benelli C, Gatteschi D, Zanchini C (1980) Electron spin resonance spectra of low-symmetry high-spin cobalt(II) complexes. 4. Tetragonal-octahedral dichlorotetrakis(pyridine)- and dichlorotetrakis(pyrazole)cobalt(II). Inorg Chem 19:1301–1304
46. Benelli C, Gatteschi D (1982) ESR spectra of low-symmetry high-spin cobalt(II) complexes. 10. Five-coordinated trigonal-bipyramidal complexes. Inorg Chem 21:1788–1790
47. Tierney DL, Rocklin AM, Lipscomb JD, Que L Jr, Hoffman BM (2005) ENDOR studies of the ligation and structure of the non-heme iron site in ACC oxidase. J Am Chem Soc 127:7005–7013
48. Miller JC, Abernathy SM, Lohr LL, Sharp RR (2000) NMR paramagnetic relaxation enhancement: ZFS-limit behavior for S = 3/2. J Phys Chem A 104:9481–9488
49. Makinen MW, Kuo LC, Yim MB, Wells GB, Fukuyama JM, Kim JE (1985) Ground term splitting of high-spin cobalt^{2+} as a probe of coordination structure. 1. Dependence of the splitting on coordination geometry. J Am Chem Soc 107:5245–5255
50. Yim MB, Kuo LC, Makinen MW (1982) Determination of the zero-field splitting constants of high-spin metalloproteins by a continuous wave microwave saturation technique. J Magn Reson 46:247–256
51. Orbach R (1961) The theory of spin-lattice relaxation in paramagnetic salts. Proc Phys Soc Lond 77:821–826
52. Daumann LJ, Comba P, Larrabee JA, Schenk G, Stranger R, Cavigliasso G, Gahan LR (2013) Synthesis, magnetic properties, and phosphoesterase activity of dinuclear cobalt(II) complexes. Inorg Chem 52:2029–2043
53. Daumann LJ, Larrabee JA, Comba P, Schenk G, Gahan LR (2013) Dinuclear cobalt(II) complexes as metallo-β-lactamase mimics. Eur J Inorg Chem 17:3082–3089
54. Vallejo J, Fortea-Perez FR, Pardo E, Benmansour S, Castro I, Krzystek J, Armentanoc D, Cano J (2016) Guest-dependent single-ion magnet behaviour in a cobalt(II) metal–organic framework. Chem Sci 7:2286–2293
55. Baum RR, Myers WK, Greer SM, Breece RM, Tierney DL (2016) The original Co(II) heteroscorpionates revisited: on the EPR of pseudotetrahedral Co(II). Eur J Inorg Chem 2016:2641–2647
56. Jesson JP, Trofimenko S, Eaton DR (1967) Spectra and structure of some transition metal poly(1-pyrazolyl)borates. J Am Chem Soc 89:3148–3158
57. Guggenberger LJ, Prewitt CT, Meakin P, Trofimenko S, Jesson JP (1973) Crystal structure and single-crystal electron paramagnetic resonance data for bis[dihydrobis(1-pyrazolyl)borato] cobalt(II). Inorg Chem 12:508–515
58. Marts AR, Greer SM, Whitehead DR, Woodruff TM, Breece RM, Shim SW, Oseback SN, Papish ET, Jacobsen FE, Cohen SM, Tierney DL (2011) Dual mode EPR studies of a kramers ion: high-spin Co(II) in 4-, 5- and 6-coordination. Appl Magn Reson 40:501–511

59. Hendrich MP, Debrunner PG (1989) Integer spin electron paramagnetic resonance of iron proteins. Biophys J 56:489–506
60. Oseback SN, Shim SW, Kumar M, Greer SM, Gardner SR, Lemar KM, DeGregory P, Papish ET, Tierney DL, Zeller M, Yap GPA (2012) Crowded bis ligand complexes of $Ttz^{Ph,Me}$ with first row transition metals rearrange due to ligand field effects: structural and electronic characterization ($Ttz^{Ph,Me}$ = tris(3-phenyl-5-methyl-1,2,4-triazolyl)borate). Dalton Trans 41:2774–2787
61. Grubel K, Marts AR, Greer SM, Tierney DL, Allpress CJ, Anderson SN, Laughlin BJ, Smith RC, Arif AM, Berreau LM (2012) Photoinitiated dioxygenase-type reactivity of open-shell 3d divalent metal flavonolato complexes. Eur J Inorg Chem 2012:4750–4757
62. Kang PC, Eaton GR, Eaton SS (1994) Pulsed electron paramagnetic resonance of high-spin cobalt(II) complexes. Inorg Chem 33:3660–3665

EPR Spectroscopy on Mononuclear Molybdenum-Containing Enzymes

Luisa B. Maia, Isabel Moura, and José J.G. Moura

Abstract The biological relevance of molybdenum was demonstrated in the early 1950s-1960s, by Bray, Beinert, Lowe, Massey, Palmer, Ehrenberg, Pettersson, Vänngård, Hanson and others, with ground-breaking studies performed, precisely, by electron paramagnetic resonance (EPR) spectroscopy. Those earlier studies, aimed to investigate the mammalian xanthine oxidase and avian sulfite oxidase enzymes, demonstrated the surprising biological reduction of molybdenum to the paramagnetic Mo^{5+}. Since then, EPR spectroscopy, alongside with other spectroscopic methods and X-ray crystallography, has contributed to our present detailed knowledge about the active site structures, catalytic mechanisms and structure/activity relationships of the molybdenum-containing enzymes.

This Chapter will provide a perspective on the contribution that EPR spectroscopy has made to some selected systems. After a brief overview on molybdoenzymes, the Chapter will be focused on the EPR studies of mammalian xanthine oxidase, with a brief account on the prokaryotic aldehyde oxidoreductase, nicotinate dehydrogenase and carbon monoxide dehydrogenase, vertebrate sulfite oxidase, and prokaryotic formate dehydrogenases and nitrate reductases.

Keywords EPR • Magnetic interactions • Molybdenum • Enzymes • Xanthine oxidase/xanthine dehydrogenase • Aldehyde oxidase • Aldehyde oxidoreductase • Nicotinate dehydrogenase • Carbon monoxide dehydrogenase • Sulfite oxidase • Nitrate reductases • Formate dehydrogenases • Dimethylsulfoxide reductas

Abbreviations

AO	Mammalian aldehyde oxidase
AOR	Bacterial molybdenum-containing aldehyde oxidoreductases
CW EPR	Continuous wave electron paramagnetic resonance spectroscopy

L.B. Maia (✉) • I. Moura • J.J.G. Moura (✉)
UCIBIO, REQUIMTE, Departamento de Química, Faculdade de Ciências e Tecnologia, Universidade Nova de Lisboa, 2829-516 Caparica, Portugal
e-mail: luisa.maia@fct.unl.pt; isabelmoura@fct.unl.pt; jose.moura@fct.unl.pt

G. Hanson, L. Berliner (eds.), *Future Directions in Metalloprotein and Metalloenzyme Research*, Biological Magnetic Resonance 33,
DOI 10.1007/978-3-319-59100-1_4

DMS	Dimethylsulfide
DMSO	Dimethylsulfoxide
DMSOR	Dimethylsulfoxide reductase
ENDOR	Electron nuclear double resonance spectroscopy
EPR	Electron paramagnetic resonance spectroscopy
FDH	Formate dehydrogenase (all types of formate dehydrogenase enzymes)
FDH-H	*E. coli* formate dehydrogenase H, from the formate-hydrogen lyase system
FDH-N	*E. coli* formate dehydrogenase N, from the anaerobic nitrate-formate respiratory pathway
Fe/S	Iron-sulfur centre
FYX (FYX-051)	4-[5-pyridin-4-yl-1H-[1,2,4]triazol-3-yl]pyridine-2-carbonitrile
mARC	Mitochondrial amidoxime reducing component
MOSC	From molybdenum cofactor sulfurase C-terminal domain (proteins involved in pyranopterin cofactor biosynthesis)
NaR	Nitrate reductase (all types of nitrate reductase enzymes, prokaryotic and eukaryotic ones)
NaRGHI	Respiratory nitrate reductase (prokaryotic), after the name of the encoding genes, *narG*, *H*, and *I*
NaRZWV	Respiratory nitrate reductase (prokaryotic), after the name of the encoding genes, *narZ*, *W*, and *V*
SO	Sulfite oxidase
XAS	X-ray absorption spectroscopy
XO	Xanthine oxidase

Introduction

The biological relevance of molybdenum was demonstrated in the early 1950s–1960s, by Bray, Beinert, Lowe, Massey, Palmer, Ehrenberg, Pettersson, Vänngård, Hanson and others, with ground-breaking studies performed, precisely, by *electron paramagnetic resonance* (*EPR*) *spectroscopy*. Those earlier studies, aimed to investigate the mammalian xanthine oxidase and avian sulfite oxidase enzymes, demonstrated the surprisingly biological reduction of molybdenum to the paramagnetic Mo^{5+}. Since then, EPR spectroscopy, alongside with other spectroscopic methods and X-ray crystallography, has contributed to our present detailed knowledge about the active site structures, catalytic mechanisms and structure/activity relationships of the molybdenum-containing enzymes. The great majority of the studies have employed continuous wave (CW) EPR, but advanced EPR-related methods were also decisive.

Molybdenum, $_{42}Mo$, is a transition metal element, belonging to the sixth group of the "d-block" of the Periodic Table, with electronic configurations [Kr] $4d^5\ 5s^1$. The paramagnetic Mo^{5+}, [Kr] $4d^1$, formed during the normal enzyme catalytic cycle or produced artificially, gives rise to characteristic d^1, $S = 1/2$, signals, with $\Delta g < 0.4$ [1]. Although the three values of g were expected to be lower than two, this is frequently not the case, as will be described (Table 1). In addition, the hyperfine structure arising from the molybdenum nucleus itself ($I = 5/2$, from ^{95}Mo and ^{97}Mo, naturally present in ca 16 and 9%, respectively) can be observed.

The relatively narrow lines of Mo^{5+} signals (small linewidths) allow the observation of weak hyperfine interactions with atoms in the close proximity of the molybdenum atom (although the IUPAC recommended denomination for this interaction type is "superhyperfine interaction", in opposition to the parent nucleus hyperfine interaction, herein, for simplicity, "hyperfine" will be used). This feature has been explored to probe the structure of the molybdenum first and second coordination spheres, in most cases using enzymes and/or substrates/inhibitors labelled with ^2H ($I = 1$), ^{13}C ($I = 1/2$), ^{15}N ($I = 1/2$), ^{17}O ($I = 5/2$), ^{33}S ($I = 3/2$) and others.

Given the large number of molybdoenzymes, this chapter does not intended to be exhaustive. Instead, to restrict the information presented to a manageable size, this Chapter will provide only a perspective on the contribution that EPR spectroscopy has made to some selected systems. Hence, after a brief overview on molybdoenzymes (section "An Overview on Molybdenum-Containing Enzymes"), the Chapter will be focused on the EPR studies of mammalian xanthine oxidase, with a brief account on the prokaryotic aldehyde oxidoreductase, nicotinate dehydrogenase and carbon monoxide dehydrogenase (section "Xanthine Oxidase Family"), vertebrate sulfite oxidase (section "Sulfite Oxidase Family"), and prokaryotic formate dehydrogenases and nitrate reductases (section "Dimethylsulfoxide Reductase Family") (Table 1).

An Overview on Molybdenum-Containing Enzymes

Molybdenum is essential to most organisms [2, 3], from archaea and bacteria to higher plants and mammals [4–11]. Actually, it is relevant for all life on Earth. Molybdenum is central to the nitrogen biogeochemical cycle, where it is mandatory for the atmospheric dinitrogen fixation (reduction) into ammonium (nitrogenase). The recent "nitrogen-to-molybdenum bio-to-inorganic bridge hypothesis" defend that the molybdenum scarcity in the Early Earth (ca 1800 Myr ago) could have delayed the evolutionary path of eukaryotes for ca 2000 Myr, by limiting the rate of dinitrogen fixation and, thus, the availability of fixed nitrogen for the early organisms [11–17]. In addition, the nitrate reduction to nitrite (nitrate reductases) and nitrite oxidation to nitrate (nitrite oxidoreductases) also depend on this metal. Molybdenum has also been suggested to be involved in the nitrite reduction to nitric oxide for signalling and survival of mammalian cells under challenging conditions [18–23]. Molybdenum is also involved in the carbon cycle, where it is used to fix (reduced) carbon dioxide into formate (formate dehydrogenases [24–26]) or interconvert aldehydes and carboxylic acids, produce plant hormones or participate in the mammalian xenobiotic metabolism (plant [27–29] and mammalian aldehyde oxidases [30–37]) or in the purine catabolism (xanthine hydroxylation to urate by

Table 1 Representative X-band EPR signals form mononuclear molybdenum-containing enzymes

Signal	$g_{1,2,3}$	$A_{1,2,3}$ (H) (MHz)[a]
Xanthine oxidase family		
Xanthine oxidase		
"Very rapid"	2.025, 1.955, 1.949	–
"Alloxanthine"	2.028, 1.959, 1.944	–
"Rapid type 1"	1.989, 1.969, 1.965	36.2, 38.3, 38.5 11.1, 8.3, 5.5
"Rapid type 2"	1.989, 1.969, 1.965	39.0, 42.0, 45.3 27.9, 29.4, 37.5
"Slow"	1.972, 1.967, 1.955	44.7, 44.4, 42.6 3.9, 4.5, 6.3
"Inhibited"	1.991, 1.977, 1.951	12.3, 10.8, 15.3
"Desulfo-inhibited"	1.980, 1.973, 1.967	–
"Arsenite"	1.973, 1.972, 1.926	–
"Mercurial"	1.969, 1.958, 1.943	–
Desulfovibrio gigas aldehyde oxidoreductas		
"Rapid type 2"	1.988, 1.970, 1.964	32.0, 45.8, 34.6 31.7, 17.4, 25.6
"Slow"	1.971, 1968, 1.958	46.4, 44.1, 39.5
"Arsenite"	1.979, 1.972, 1.922	–
Eubacterium barkeri nicotinate dehydrogenase		
"Very rapid" like	2.067, 1.982, 1.974	–
Oligotropha carboxidovorans carbon monoxide dehydrogenase		
"Mo/Cu"	2.001, 1.960, 1.955	–[b]
Sulfite oxidase family		
Sulfite oxidase		
"Low pH"	2.004, 1.972, 1.966	23.8, 22.1, 35.8 4, 6, 4
"High pH"	1.987, 1.964, 1.953	–[c]
"Phosphate-inhibited"	1.992, 1.969, 1.974	–
Dimethylsulfoxide reductase family		
Formate dehydrogenases		
Escherichia coli formate dehydrogenase H	2.094, 2.001, 1.990	7.5, 18.9, 20.9
Desulfovibrio desulfuricans formate dehydrogenase	2.012, 1.996, 1.985	23.1, 29.9, 27.8 35.1, nd, nd
Ralstonia eutropha NAD-dependent formate dehydrogenase	2.009, 2.001, 1.992	18, 21, 18
D. alaskensis molybdenum-containg formate dehydrogenase	1.971, 1.968, 1.959	44.2, 44.1, 43.9
Periplasmatic nitrate reductase		
Paracoccus pantothrophus periplasmatic nitrate reductase		
"Low *g*-unsplit"	1.997, 1.962, 1.959	–
"Low *g*-split"	1.996, 1.969, 1.961	36.3, 37.5, 42.0
"Very high *g*"	2.022, 1.999, 1.994	20.9, 20.7, 18.4

(continued)

Table 1 (continued)

Signal	$g_{1,2,3}$	$A_{1,2,3}$ (H) (MHz)[a]
"High *g*"	1.999, 1.990, 1.981	17.9, 14.5, 13.9 8.4, nd, nd
"High *g*-nitrate"	1.999, 1.989, 1.982	17.9, 12.0, 12.8 9.0, nd, nd
Desulfovibrio desulfuricans periplasmatic nitrate reductase		
"High *g*-nitrate"	2.000, 1.990, 1.981	12.9, 13.9, 12.8
"High *g*-turnover"	1.999, 1.992, 1.982	16.2, 18.1, 15.3 16.2, 18.1, 15.3
Assimilatory nitrate reductase		
Synechococcus sp. PCC 7942 assimilatory nitrate reductase		
"Very high *g*"	2.023, 1.998, 1.993	–
"high *g*"	1.997, 1.990, 1.982	–
Respiratory nitrate reductases		
Escherichia coli respiratory nitrate reductase		
"Low pH"	2.001, 1.986, 1.964	31.7, 23.6, 24.7
"High pH"	1.988, 1.981, 1.962	10.6, 8.9, 9.1
Paracoccus pantothrophus respiratory nitrate reductase		
"Low pH"	2.007, 1.987, 1.970	nd
"High pH"	1.990, 1.989, 1.967	nd
M. hydrocarbonoclasticus respiratory nitrate reductase		
"Low pH-nitrate"	1.996, 1.982, 1979	nd
"High pH-nitrtate"	2.002, 1.987, 1968	39.2, 30.6, 30.3

[a]*nd* not detected

[b]$A_{1,2,3}$(Cu) = 117, 164, 132 MHz

[c]An ENDOR study demonstrated the presence of a solvent-exchangeable, strongly and anisotropically coupled proton in this signal [253–255]

xanthine oxidoreductase [38–44]). The role of molybdenum is also extended to the sulfur cycle, where it is used in the respiratory oxidation of inorganic sulfur compounds (sulfite-oxidising enzymes) [45–48]. It is also critical to the catabolism of sulfur-containing compounds, being vital for human cells that must eliminate (oxidise) the toxic sulfite to survive (sulfite oxidase) [49–54].

Presently, more than 50 molybdenum-containing enzymes are known, many of which have been biochemically and structurally characterised, and the number is increasing every year, with several more being foreseen to be "discovered" in a near future based on genomic analyses. Noteworthy, the great majority of the molybdoenzymes are prokaryotic, whereas only a restricted number of molybdoenzymes are found in eukaryotes.

Organisms use molybdenum in the active site of enzymes that catalyse (almost exclusively) oxidation/reduction reactions at carbon, nitrogen and sulfur atoms of key metabolites, most of which involve the transfer of one oxygen atom [4–11]. The chemical properties of molybdenum are perfectly suitable for this "Redox

Biochemistry" [55]: it is redox-active under physiological conditions, where its oxidation state can range from 6+, 5+ and 4+, and even 3+, in nitrogenase [56, 57]; it can have a very versatile first coordination sphere; its chemistry is dominated by the formation of oxides and sulfides, where its strong tendency to bind oxo groups is balanced by its ability to easily lose a single oxygen atom. This feature makes molybdenum centres excellent "oxygen atom exchangers" [58–66], as long as the thermodynamics of the reaction of "oxygen exchange" is favourable (Eqs. (1) and (2)) [59, 64, 67], what has led to the "oxo transfer hypothesis" coined by Holm and others in the 1980s. Organisms explore this rich chemistry to carry out different oxotransfer reactions (see Eqs. (5), (6), (7), (8), (9), (10), and (11) in section "EPR Studies of Molybdoenzymes"), where one oxygen atom is transferred from water to product *-oxygen atom insertion* (Eq. (3))- or from substrate to water *-oxygen atom abstraction* (Eq. (4)). These reactions involve a net exchange of two electrons, with the molybdenum atom cycling between Mo^{6+} and Mo^{4+} (Eqs. (3) and (4)), and, most importantly, with the metal being the direct oxygen atom donor or acceptor (Eqs. (1) and (2)) [4–11, 31, 35, 38, 41, 42, 44]. Subsequently, the initial metal oxidation state is regenerated, in most of the cases, by two one-electron oxidation/reduction reactions ($Mo^{6+} \leftrightarrow Mo^{5+} \leftrightarrow Mo^{4+}$) with other redox-active cofactors within the enzyme (iron-sulfur (Fe/S) centres, haems, flavins). Noteworthy, some molybdoenzymes are able to catalyse both oxygen atom insertion and abstraction during the same catalytic cycle [11, 18–23, 68].

$$Mo-O+X \rightarrow Mo+X-O \quad (1)$$

$$Mo+X-O \rightarrow Mo-O+X \quad (2)$$

$$Mo^{6+}+Q+H_2O \rightarrow Mo^{4+}+Q-O+2H^+ \quad (3)$$

$$Mo^{4+}+R-O+2H^+ \rightarrow Mo^{6+}+R+H_2O \quad (4)$$

The versatile chemistry of molybdenum allows it to also catalyse reactions of sulfur atom transfer [69–73], hydrogen atom transfer (Eq. (12), in section "EPR Studies of Molybdoenzymes") and even a non-redox hydration reaction (the acetylene hydratase-catalysed hydration of acetylene to acetaldehyde).

Structurally, molybdenum is found in the enzymes active site in a mononuclear form [4–11], except in nitrogenase, where it is present in the unique heteronuclear [$MoFe_7S_9C$], and in a few other cases[1]. In the mononuclear molybdenum centres, one molybdenum atom is coordinated by the *cis*-dithiolene (–S–C=C–S–) group of one or two pyranopterin cofactor molecules (Fig. 1a) and by oxygen and/or sulfur and/or selenium atoms in a diversity of arrangements. Based on the *metal center structure*,

[1] The carbon monoxide dehydrogenase from *Oligotropha carboxidovorans* or *Hydrogenophaga pseudoflava*, with its unique binuclear Mo/Cu cofactor (Mo-S-Cu-S(Cys)) [74–78], as well as a few other heteronuclear centres of proteins whose physiological function is not yet fully understood [79–84], are exceptions to the mononuclear presence of molybdenum. In spite of that, the carbon monoxide dehydrogenase is classified as a member of the xanthine oxidase family.

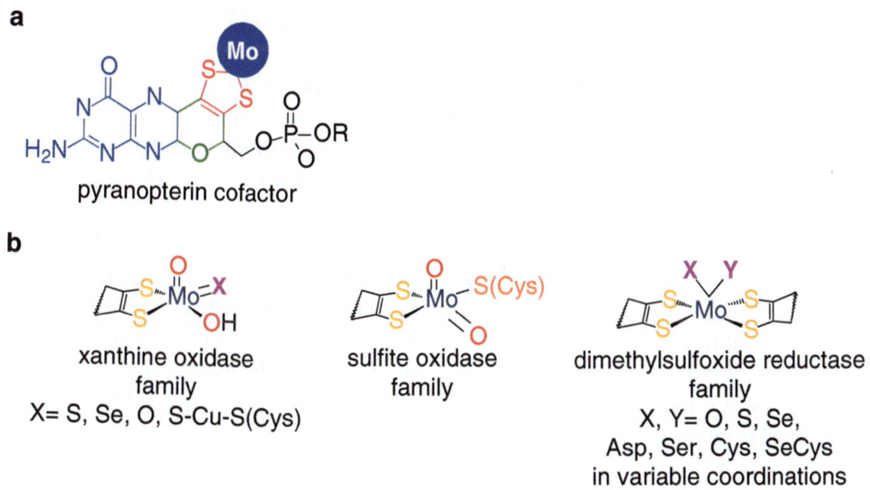

Fig. 1 Structures of the pyranopterin cofactor (**a**) and of the active site of molybdenum-containing enzymes. (**a**) The pyranopterin cofactor molecule is formed by pyrano(*green*)-pterin(*blue*)-dithiolene(*red*)-methylphosphate(*black*) moieties. The dithiolene (–S–C=C–S–) group forms a five-membered ene-1,2-dithiolate chelate ring with the molybdenum/tungsten atom. In eukaryotes, the cofactor is found in the simplest monophosphate form (R is a hydrogen atom), while in prokaryotes it is most often found esterified with several nucleotides (R can be one cytidine monophosphate, guanosine monophosphate or adenosine monophosphate). (**b**) Structures of the molybdenum centres of the three families of molybdoenzymes in the oxidised form. For simplicity, only the *cis*-dithiolene group of the pyranopterin cofactor is represented. Adapted with permission from [21]

the mononuclear molybdoenzymes are organised into three families, denominated after one benchmark enzyme (Fig. 1b) [4], which will be described in the following sections: xanthine oxidase (XO; section "Xanthine Oxidase Family") family, sulfite oxidase (SO; section "Sulfite Oxidase Family") family and dimethylsulfoxide reductase (DMSOR; section "Dimethylsulfoxide Reductase Family") family.

EPR Studies of Molybdoenzymes

Xanthine Oxidase Family

The Enzymes

The active site of the XO family enzymes (in its oxidised form; Fig. 1b) has a molybdenum atom coordinated in a distorted square-pyramidal geometry by an apical oxo group (Mo=O) and, in the equatorial plane, by the two sulfur atoms of the *cis*-dithiolene group of one pyranopterin cofactor molecule, one catalytically labile -OH group (in almost all enzymes) plus one terminal sulfo (Mo=S; in most enzymes) or seleno (Mo=Se) or oxo (Mo=O) or Mo-S-Cu-S(Cys) moiety (see note 1) [4–11, 31, 35]. This family comprises the prototype mammalian (bovine milk) XO (Eq. (5)) and other enzymes such as mammalian aldehyde oxidases (AO), *Desulfovibrio* aldehyde oxidoreductases (AOR), *Eubacterium barkeri* nicotinate dehydrogenase, *Oligotropha carboxidovorans* carbon monoxide dehydrogenase, *Pseudomonas putida* quinoline 2-oxidoreductase or *Thauera aromatica* 4-hydroxy-benzoyl-CoA reductase.

Structurally, mammalian XO is an homodimeric enzyme, with each monomer holding two [2Fe-2S] centres and one FAD, besides one molybdenum centre [38, 41, 42, 44, 85, 86]. The four redox-active centres are aligned in an almost linear fashion, defining an intramolecular electron transfer pathway that delivers electrons from the molybdenum centre to the FAD, the sites where the hydroxylation (Eq. (6)) and dioxygen reduction (Eq. (7)) take place, respectively, with Fe/S centres intermediating the electron transfer between the two.

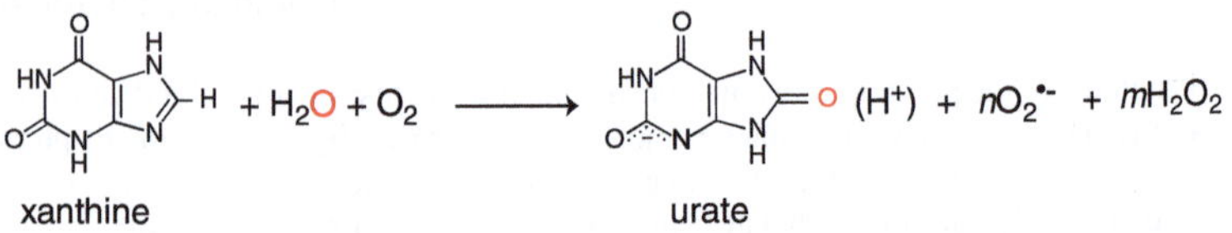

(5)

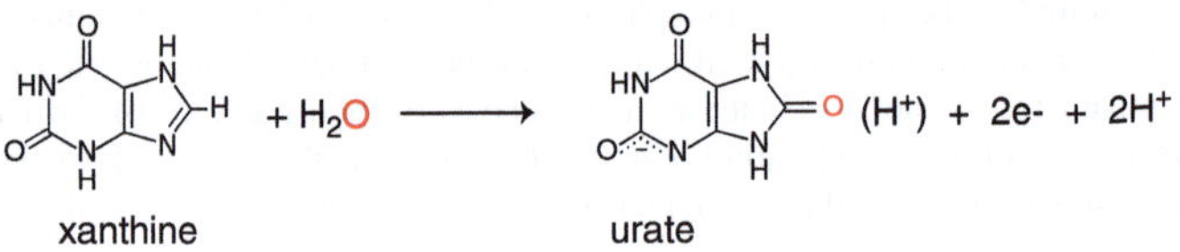

(6)

$$O_2 + 2e^- + 2H^+ \rightarrow nO_2^{\cdot -} + mH_2O_2 \tag{7}$$

XO family enzymes typically catalyse the hydroxylation of a C–H moiety in aromatic heterocyclic compounds and aldehydes [31, 35, 38, 41, 42, 44]. For this, the enzymes molybdenum centre has to promote the cleavage of a C–H bond and the formation of a novel C–O bond, as the XO-catalysed reaction of xanthine hydroxylation to urate illustrates (Eq. (6)). The molecular mechanism of XO-catalysed hydroxylation reaction is presently well established and believed to be essentially similar in other members of this family, namely in AO and AOR enzymes (Fig. 2) [31, 35, 38, 41, 42, 44, 85–89]: (1) the catalysis is initiated with the activation of the molybdenum catalytically labile hydroxyl group (Mo-OH) by a neighbouring conserved deprotonated glutamate residue, to form an Mo^{6+}-O^- core (base-assisted catalysis); (2) the now deprotonated oxygen undertakes nucleophilic attack on the carbon atom to be hydroxylated, with the simultaneous transfer of hydride from substrate to the essential sulfo group (Mo=S → Mo–SH), resulting in the formation of a covalent intermediate, Mo^{4+}-O-C-R(-SH) (where R represents the remainder of the substrate molecule); (3) hydroxide/water (from solvent) then displaces the hydroxylated product from the molybdenum coordination sphere to yield a Mo^{4+}-$OH_{(2)}$(-SH) core (oxidation half-reaction; Eq. (6)); (4) the two electrons transferred from the substrate to the molybdenum ($Mo^{6+} \rightarrow Mo^{4+}$) during the

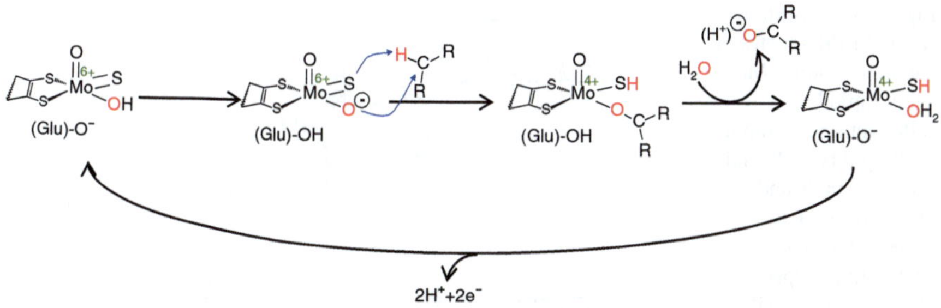

Fig. 2 Mechanism of xanthine oxidase-catalysed hydroxylation reaction. See text for details

oxidative half-reaction are rapidly transferred, via the Fe/S centres, to the FAD (Mo → Fe/S → FAD), where the reduction of dioxygen takes place (reductive half-reaction; Eq. (7)); (5) in the now oxidised molybdenum centre (Mo^{4+} → Mo^{6+}), the sulfo group is deprotonated and the initial Mo^{6+}-OH(=S) core is regenerated.

The critical role of the terminal sulfo group, Mo=S (Fig. 2), in catalysis is revealed when, after its removal, the enzymes hydroxylation activity is completely abolished [90, 91]. The chemical removal of the sulfo group is achieved through reaction with cyanide, which abstracts the sulfur atom in the form of thiocyanide and leaves an oxo group, Mo=O, in its place. The modified enzyme form is denominated "desulfo", in opposition to the "sulfo" form of native enzyme. Noteworthy, the desulfo form can be converted into the sulfo form (Mo=O → Mo=S) by sulfuration through reaction with sulfide (Na_2S). "Desulfo" XO molecules occur naturally in cells and are the result of inefficient in vivo sulfuration [92–94].

Two other features of the XO reaction mechanism have to be here highlighted. First, water is the ultimate source of the oxygen atom incorporated into the hydroxylated product (Eq. (6); Fig. 2), as is characteristic of molybdoenzymes, with dioxygen being *only* the oxidising substrate [95–97] (Eq. (7)). The direct oxygen donor is the catalytically labile equatorial Mo-OH group of the molybdenum center; this group (this oxygen atom) is subsequently regenerated by a water molecule from the solvent, in the end of each catalytic cycle. Second, the hydroxylation reaction occurs through one two-electron reduction of the molybdenum centre (Mo^{6+} → Mo^{4+}) [98] and the intramolecular electron transfer to other redox centres within the enzyme (Mo → Fe/S → FAD, in the case of XO) is an integral aspect of the catalysis. Hence, the Mo^{5+} species observed by EPR spectroscopy related to hydroxylation reactions are formed through the oxidation of the Mo^{4+} species.

EPR Studies of Mammalian Xanthine Oxidase

Being one of the most studied molybdoenzymes, our present knowledge about XO structure and reaction mechanism, together with the information gathered from the study of model compounds [66, 99–102], allows us to discuss its EPR signals in

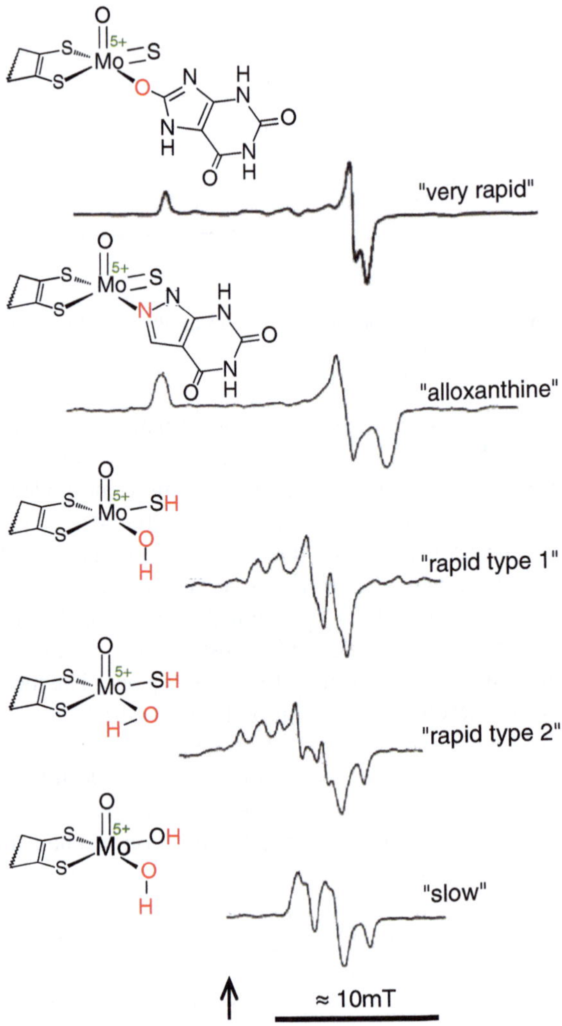

Fig. 3 Representative X-band EPR spectra of xanthine oxidase Mo^{5+} "very rapid", "alloxanthine", "rapid type 1", "rapid type 2" and "slow" signals and the respective structures proposed for each signal-giving species. "Very rapid" signal obtained with xanthine; adapted with permission from [104], copyright (1988) American Chemical Society. "Alloxanthine" signal; adapted with permission from [112]. "Rapid type 1" signal obtained with purine; adapted with permission from [145], copyright (1982) American Chemical Society. "Rapid type 2" signal obtained with dithionite in the presence of borate; adapted with permission from [145], copyright (1982) American Chemical Society. "Slow" signal obtained with dithionite; adapted with permission from [145], copyright (1982) American Chemical Society. The *arrow* indicates the position of $g = 2.0037$

detail and to draw a clear picture of the structure of each signal-giving species—as will be discussed in the following sections (Figs. 3 and 4).

The biological relevance of molybdenum was unambiguously and definitively demonstrated by Bray and Meriwether in 1966 [103] using XO purified form milk of cows that had been injected with ^{95}Mo-labelled molybdate (e.g., $A_{1,2,3}(^{95}Mo) =$ 133, 54.7, 57.3 MHz, with xanthine, in the "very rapid" signal [104]). Those earlier studies demonstrated the surprising presence and catalytic role of molybdenum in a mammal [105, 106].

The XO Mo^{5+} can give rise to different EPR signals, depending on the enzyme form ("sulfo", "desulfo", inhibited), the compound used to reduce the enzyme (purine derivates, aldehydes, artificial reductants, etc) and on the time of reduction.

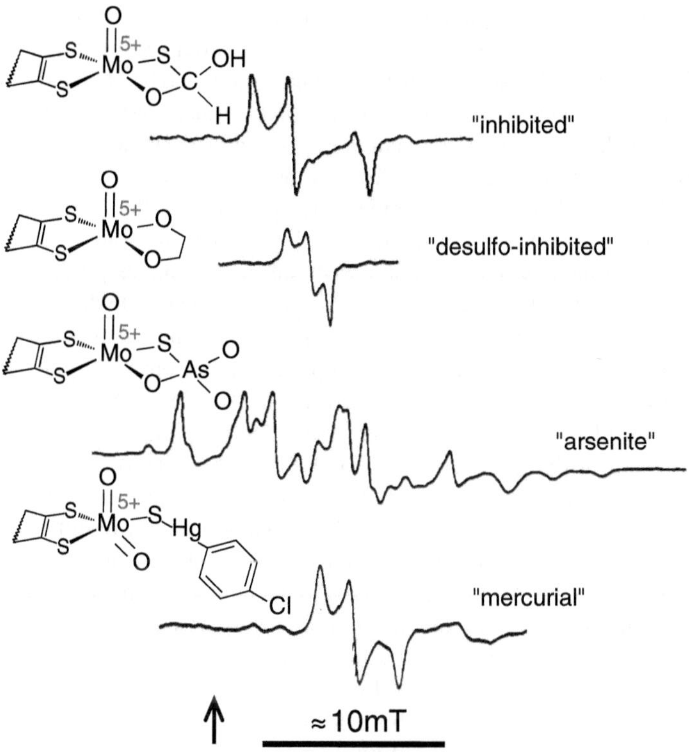

Fig. 4 Representative X-band EPR spectra of xanthine oxidase Mo^{5+} "inhibited", "desulfo-inhibited", "arsenite" and "mercurial" signals and the respective structures proposed for each signal-giving species. "Inhibited" signal obtained with formaldehyde; adapted with permission from [158]. "Desulfo-inhibited" signal obtained with ethylene-glycol; adapted with permission from [145], copyright (1982) American Chemical Society. "Arsenite" signal obtained with arsenite-treated and dithionite-reduced enzyme; adapted with permission from [165]; copyright (1983) American Chemical Society. "Mercurial" signal obtained with *p*-chloro-mercuri-benzoate; adapted with permission from [168]; copyright (1983) American Chemical Society. The *arrow* indicates the position of $g = 2.0037$

The signals were originally named by Bray and Vänngård [107], based on the relative rate of their formation and decay in the course of the reaction of enzyme with substrate, as "very rapid" ($t_{1/2} \approx 10^{-2}$ s), "rapid" ($t_{1/2} \approx 10^{-1}$ s), "slow" ($t_{1/2} \approx 10^{3}$ s) and "inhibited" ($t_{1/2} \approx 10^{6}$ s) signals (Figs. 3 and 4). Later, the observation of other signals led to the introduction of the "desulfo-inhibited", "arsenite" or "mercurial" signals.

The "Very Rapid" Signal

The "very rapid", originated from XO molecules in the native, sulfo, form, is a nearly axial signal ($g_1 \gg g_2 \approx g_3$), with $g_1 > 2$ ($g_{1,2,3} = 2.025, 1.955, 1.949$, with xanthine [108]), that does not display hyperfine structure due to coupled protons

(Fig. 3) [108, 109]. This signal was observed for the first time in 1964, by Palmer et al. [105], in the presence of an excess of xanthine, after a short reaction period (8 ms). Subsequent studies showed that the signal could also be obtained in the presence of sub-stoichiometric amounts of xanthine (either anaerobically or aerobically), as well as, with hypoxanthine, 1-methyl-xanthine, 6-methyl-purine, 2-hydroxy-6-methyl-purine, 8-hydroxy-6-methyl-purine or 2,4-dihydroxy-pteridine, being its formation favoured by high pH values [106, 109–112]. The species that gives rise to the "very rapid" signal is thought for long to be a catalytic intermediate, formed by oxidation of a Mo^{4+}-substrate complex [107, 109, 113].

Studies with $^{13}C8$-labelled xantina (the position that is hydroxylated by the enzyme), in ^{17}O-labelled water (the oxygen that is ultimately incorporated in urate), resulted in a "very rapid" signal splitted approximately isotropically by both nuclei ($A_{1,2,3}(^{13}C)$ = 11.1, 7.6, 7.6 MHz [108] and $A_{1,2,3}(^{17}O)$ = 38.0, 38.3, 37.1 MHz [104, 108, 109, 114, 115]). The strong isotropic hyperfine coupling to ^{17}O was taken as the main evidence for the presence of a Mo-$O_{non\text{-}terminal}$ bond (Mo-O-R) in the signal-giving species (in opposition to the small and quite anisotropic interaction ($A_{1,2,3}(^{17}O)$ = 8.6, 8.6, 1.9 MHz) observed in model compounds with a terminal oxo group, Mo=O [116]). In its turn, the splitting induced by ^{13}C showed that the xanthine C8 is located within magnetic contact to the molybdenum atom and, although the small interaction value could arise from an equatorial Mo-C8(xanthine) complex (that is, from an equatorial ligand σ-bonded to the molybdenum, with the ligand laying on a node of the ground-state orbital), that hypothesis is difficult to rationalise mechanistically and to reconcile with the above Mo-O-R interpretation. Hence, the signal was interpreted as probably arising form of a Mo^{5+}-O-C8(xanthine) complex.

Noteworthy, alloxanthine originates an EPR Mo^{5+} signal very similar to the "very rapid", also nearly axial, with $g_1 > 2$ (2.028, 1.959 and 1.944) and with no coupled protons (Fig. 3) [112, 114]. Alloxanthine, a xanthine isomer that holds one nitrogen atom (N2) in place of the xanthine C8, is a powerful inhibitor of the hydroxylation activity of XO [117]. In the presence of alloxanthine, the XO Mo^{5+} interacts isotropically with a single nitrogen atom (presumably the alloxanthine N2; $a_{1,2,3}(^{14}N)$=9.9, 9.6, 8.7 MHz), but not with ^{17}O, probably through a Mo^{5+}-N2(alloxanthine) complex [112].

The use of ^{95}Mo- and ^{97}Mo-enriched XO (^{95}Mo, $I = 5/2$, no significant nuclear electric quadrupole; ^{97}Mo, $I = 5/2$, strong nuclear electric quadrupole) showed a strong anisotropic hyperfine coupling, as well as, nuclear electric quadrupole interaction, with $A_{1,2,3}(^{95}Mo)$ = 142, 60.1, 63.4 MHz, with non-coincident $\boldsymbol{g}$ and $\boldsymbol{A}$ tensors axes (with Euler angles of $\alpha = 7°$, $\beta = 42°$ and $\gamma = 0°$) and $P_{1,2,3}(^{97}Mo)$ = 4.0, −5.5, 1.5 MHz, with coincident $\boldsymbol{P}$ and $\boldsymbol{A}$ tensors axes (using 2-hydroxy-6-methyl-purine) [104]. The highly anisotropic $\boldsymbol{P}$, with the largest principal axis pointing along A_2, was most consistent with a MoOS structure for the "very rapid" signal-giving species. Using density functional theory, Drew and Hanson [118] and others [119, 120] suggested that the non-coincidence angle β could provide a measure of the metal-dithiolene fold angle of the molybdenum centre (defined by the dihedral angle between the MoS_2 plane and the S_2C_2 plane of the pyranopterin molecule). Accordingly, the above mentioned angle β of 42° would correspond to a metal-dithiolene fold angle of 15° [118]. Although this value is somewhat different from the 21° observed in the crystal structure of XO complexed with 2-hydroxy-6-methyl-purine [87] or with FYX [86], it can be argued that crystallographic structures represent a weighted mean of oxidation states (Mo^{6+}, Mo^{5+}, Mo^{4+}).

When the "very rapid" signal is generated from XO molecules that were desulfurated (reaction with cyanide) and resulfurated with ^{33}S-labelled sulfide (reaction with $Na_2{}^{33}S$), a strong and anisotropic interaction is observed ($A_{1,2,3}(^{33}S) = 8.5, 76.6, 19.1$ MHz, with xanthine) [91, 104, 121, 122]. The related signal originated in the presence of alloxanthine also displays a similar coupling to ^{33}S ($A_{1,2,3}(^{33}S) = 8.5, 85.0, 19.1$ MHz) [112, 123]. The observed anisotropy suggested that the sulfur atom would be a terminal ligand, Mo=S and not Mo-S-R (contrary to the oxygen atom, that would be as Mo-O-R and not Mo=O); the absence of hyperfine coupling to protons depicted this sulfur as a Mo=S (and not Mo-S-H). Furthermore, because the equatorial ligands originate stronger interactions than the axial ligands in model compounds, the sulfur atom was suggested to be in an equatorial position [122, 124]. The magnitude of the $Mo^{5+}/^{33}S$ anisotropic interaction, unexpected for an isotope with a small nuclear g value (g_n of 0.429), suggests that the sulfur p orbital participates to a large extent in the ground-state molecular orbital of the signal-giving species, with the unpaired electron delocalised ca 38% over the sulfur [104, 109, 122] (calculated based on the expected anisotropic coupling of 156 MHz, if the unpaired electron resided entirely in a sulfur p orbital [125]). Also the unusually high value of g_1 (that makes the signal look like an axial one) was suggested to result mainly from spin-orbit coupling with filled orbitals involving this sulfur, supporting the claim that a $g > 2$ is evidence for sulfur ligation [104]. If the unpaired electron is delocalised over the sulfur, then one group exerting a strong ligand field is necessary to bring the ground-state orbital into approximately the same plane as the sulfo group [104]. In 1985, Bray and George [109] (without having any evidence for its existence, now confirmed by X-ray crystal structures) ingeniously suggested the presence of one "spectator" axial oxo group.

In summary, the structure of "very rapid" signal-giving species suggested by the EPR data is a **$Mo^{5+}(=O_{axial})(=S_{equatorial})$-$O_{equatorial}$-C(xanthine)** complex (Fig. 3). This structure was entirely confirmed, when the crystallographic structure of oxidised XO [85] and of the key intermediate in the hydroxylation reaction of FYX by XO [86] were solved. The reduced enzyme-substrate (XO-FYX) complex obtained by Okamoto et al. [86] revealed the presence of an axial oxo group and one equatorial sulfo group, unambiguously modelled as a Mo-SH, as well as the substrate molecule bound through the labile oxygen atom to the molybdenum atom. This reduced enzyme-substrate complex, in which the carbon–oxygen bond of the product was already formed, but the product remained bound to the molybdenum, can, thus, be formulated as a $Mo^{4+}(=O_{axial})(-SH_{equatorial})$-$O_{equatorial}$-C(FYX) complex. The following cleavage of the Mo^{4+}-O bond results in the hydroxylated product release; the $Mo^{4+}(=O_{axial})(-SH_{equatorial})$-$OH_{equatorial}$ complex formed is subsequently oxidised, via intramolecular electron transfer, to regenerate the initial Mo^{6+} complex. The Mo^{4+}-substrate reaction intermediate can, however, follow an alternative pathway, in which it is first one-electron oxidised to the Mo^{5+}-substrate species that gives rise to the "very rapid" signal and, only then, cleaved to release the hydroxylated product.

Accordingly, it is the balance between the rate of Mo^{4+}-substrate oxidation to Mo^{5+}-substrate and the rate of product release from Mo^{4+}-substrate that determines if a given substrate can, or not, originate a "very rapid" signal [109, 126]. When it is formed, the "very rapid" signal is originated from a catalytic intermediate [86, 107, 109, 113].

The structure of the "very rapid" signal-giving species was also probed by X-ray absorption (XAS) and electron nuclear double resonance (ENDOR) spectroscopies. Both supported the presence of a Mo=O group [121, 127] with ENDOR showing the presence of only one single ^{17}O-oxygen atom in the coordination sphere of molybdenum [128]. That earlier ENDOR study suggested the formation of a direct Mo-C(substrate) intermediate (Mo…C distance ≤ 2.4 Å) [128], but, later, Manikandan et al. in 2001 [129], assuming that the carbon undergoes a sp^2-hybridisation, suggested instead a Mo…C distance of 2.8 Å, a value similar to the one later observed in the crystallographic structure of the XO-FYX complex and consistent with a Mo-O-C(substrate) complex.

The structurally related XO-alloxanthine complex can, in a similar way, be formulated as **$Mo^{5+}(=O_{axial})(=S_{equatorial})$-$N_{equatorial}$(alloxanthine)** (Fig. 3) and this structure was also confirmed by X-ray crystallography [130].

The "Rapid" Signals

The signals of the "rapid" type, originated also from XO molecules in the native, sulfo, form, can be easily distinguished from the "very rapid" by their rhombicity ($g_1 > g_2 > g_3$), by having $g_1 < 2$ and by the presence of hyperfine structure due to coupled protons (e.g., $g_{1,2,3}$ = 1.989, 1.969, 1.965, with 1-methyl-xanthine [115, 131]) (Fig. 3) [107, 131, 132]. The "rapid" signals can be obtained apparently with all substrates, within the turnover time. Noteworthy, they can be generated with substrates/reagents that do not reduce the enzyme through the molybdenum, as is the case of NADH (that reduces the XO through the FAD) or dithionite [110, 133–141]. Yet, they are not observed in single turnover assays with purinic substrates [113, 142, 143]. Moreover, the addition of xanthine causes very little changes in the signal [107, 134, 142], showing that the presence of the substrate has little influence in the coordination sphere of pre-reduced molybdenum and, thus, that the xanthine molecule does not become bound to the pre-reduced molybdenum. In compliance, no interaction with ^{13}C-labelled purinic substrate is observable in the "rapid" signals. Hence, in contrast to the "very rapid", the "rapid" signal-giving species does dot have the substrate bound and it does not represent a productive catalytic intermediary of the XO reaction [142].

The hyperfine interactions with ^{17}O and ^{33}S are in the "rapid" signals strong anisotropic and weak isotropic respectively ($A_{av}(^{17}O)$ = 4.2, 8.3, 44.0 MHz [144–146] and $A_{1,2,3}(^{33}S)$ = 9.5, 9.9, 9.9 MHz [104, 122], using formamide). The strong anisotropic hyperfine coupling to ^{17}O was initially taken as evidence for the presence of a terminal oxo group (Mo=O and not Mo-O-R, following the same reasoning described above for the "very rapid" signal). Yet, later studies showed, instead, that the ^{17}O interaction could arise from a non-terminal oxygen, a Mo-OH group (further discussed below), in a ligand field that includes a "spectator", axial Mo=O group [97, 147]. The isotropic coupling to ^{33}S suggested a non-terminal sulfur (Mo-S-R and not Mo=S) in the signal-giving species and, contrary to the "very rapid" signal, that the unpaired electron is not delocalised over the sulfur atom. In addition, the strong and anisotropic interaction with ^{95}Mo and ^{97}Mo ($A_{1,2,3}(^{95}Mo)$ = 184, 74.0, 77.0MHz and $P_{1,2,3}(^{97}Mo)$ = 4.0, −2.0, −2.0 MHz, with coincident ***P*** and ***A*** tensors axes, but not coincident with the ***g*** axes (with Euler angles of $\alpha = 0°$, $\beta = 18°$ and $\gamma = 0°$); with formamide [104]) was also taken as evidence for a Mo-SH(-OH) structure [123, 148].

Also in contrast to the "very rapid", the "rapid" signals display hyperfine interactions with solvent-exchangeable protons and two types can be defined. In "rapid type 1", the molybdenum interacts with two non-equivalent protons (e.g., $A_{1,2,3}$ = 36.2, 38.3, 38.5 MHz and 11.1, 8.3, 5.5 MHz, with 1-methyl-xanthine [115, 131, 146]), while in "rapid type 2" it interacts with two equivalent protons (e.g., $A_{1,2,3}$ = 39.0, 42.0, 45.3 MHz and 27.9, 29.4, 37.5 MHz, with borate [122]) (Fig. 3). In the presence of xanthine, e.g., mixtures of the two types are usually obtained, but higher xanthine concentrations and lower pH values favour the "type 1" [132, 134].

Comparative studies of the "rapid type 1" signal with ^{33}S-labelled XO and 1-methyl-8-^{1}H-xanthine versus 1-methyl-8-^{2}H-xanthine, in ^{1}H versus ^{2}H-labelled water, showed that the strongest coupled proton is originated from the C8 carbon of the substrate molecule that was hydroxylated in the previous catalytic cycle and this proton was attributed to the sulfur atom of the molybdenum centre (Mo-S-H)[2] [91, 122, 131, 138, 149, 150]. The apparent incompatibility between the weak interaction with the sulfur (a molybdenum ligand) and the strong interaction with the proton (not a molybdenum ligand) can be explained by a non-linear Mo-S-H structure [122]. An angular Mo-S-H allows the sulfur atom to be located near the nodal plane of the ground-state orbital of the molybdenum (in a lower electron density site), with the hydrogen atom out of the nodal plane into a region of higher electron density. The weakly coupled proton of "rapid type 1" could reside in the oxygen or in a neighbouring protein group with which the oxygen forms an hydrogen bond [104, 144]. This interpretation was supported by the similarity of the EPR spectra of model compounds [151] and, because the "very rapid" does not show interaction with protons, the oxygen atom involved can not be the axial one.

The two strongly coupled protons in the "rapid type 2" signals have also been attributed to the Mo-SH and Mo-OH groups [114]. It has been suggested that the Mo-SH group holds a strongly coupled proton in both signal types and it is the relative position of the proton in the Mo-OH group relatively to the molybdenum that

[2] Although the ^{2}H has a nuclear magnetic moment different from zero (I = 1), in CW EPR the expected splitting ($2I + 1 = 3$) is not observable. The hyperfine interaction due to hydrogen is usually weak and the line splitting due to ^{2}H is only *ca* 15% of the splitting due to ^{1}H (calculated based on the ration of the values of nuclear g_n, which are 5.59 and 0.86 for ^{1}H and ^{2}H, respectively). Hence, in practice, in CW EPR, ^{1}H substitution by ^{2}H "removes" the previous observable split by 2 ($2I + 1/2 = 2$).

determines if that proton is weakly ("rapid type 1") or strongly ("rapid type 2") coupled. Thus, following the same line of thinking used to rationalise the weak interaction with the sulfur and the strong interaction with the proton in the "rapid type 1" signal, the stronger interaction in the "rapid type 2" is due to a different geometry of the molybdenum centre and not to a molybdenum centre with a different structure.

Although the hyperfine interactions (with protons, sulfur, oxygen and carbon) are different in "very rapid" and "rapid" signals, reflecting changes in the geometry of the molybdenum centre, the relative position of the direct molybdenum ligands must remain unchanged and the basic structure of the "very rapid" signal-giving species can be here used as template. Hence, the structure of "rapid" signal-giving species suggested by the EPR data is a $\mathbf{Mo^{5+}(=O_{axial})(-SH_{equatorial})-OH_{equatorial}}$ complex, with the Mo-OH proton being present in different orientations in "type 1" and "type 2" signals (Fig. 3).

The "Slow" Signal

The "slow" signal is originated from XO molecules in the desulfo form, where the sulfo group, Mo=S, was substituted by an oxo group, Mo=O. Although this signal-giving species is not catalytically relevant (because "desulfo" molecules do not have hydroxylase activity), it is physiologically relevant (because it is present in cells) and its study provides further relevant structural information (confirmation). The signal could be obtained apparently with any substrate/reductant provided that "desulfo" enzyme molecules are present. Swann and Bray, in 1972 [136], demonstrated with xanthine-reduced XO that the formation of the "slow" signal is dependent on the slow intermolecular electron transfer, from reduced "sulfo" enzyme molecules to oxidised "desulfo" molecules. The lower reduction potential of the "desulfo" molybdenum centre ($\approx$100 mV lower than the "sulfo" centre value [152]) further challenges its reduction [153]. Swann and Bray [136] demonstrated also that NADH (that reduces the XO through the FAD) is able to reduce both the "sulfo" and the "desulfo" enzyme molecules, leading to the simultaneous arising of "rapid" and "slow" signals. Moreover, the intensity of "rapid" and "slow" signals is proportional to the concentration of "sulfo" and "desulfo" XO molecules, respectively. Hence, the "desulfo" enzyme molecules can be reduced either by intermolecular electron transfer or by direct reduction and the co-presence of "rapid" and "slow" signals can be observed with any substrate/reductant (e.g., xanthine, NADH or dithionite [136, 141, 152–155]), as long as sufficient time is giving for the reduction to occur.

The "slow" signal is characterised by the presence of hyperfine structure due to two non-equivalent protons ($A_{1,2,3}$ = 44.7, 44.4, 42.6 MHz and 3.9, 4.5, 6.3 MHz, with dithionite) and by a lower g_{av} value ($g_{1,2,3}$ = 1.972, 1.967, 1.955), which allows it to be easily differentiated from the "rapid type 1" signal (Fig. 3) [123, 131, 138, 148, 156, 157]. The lower g_{av} value was suggested to result from a lower spin-orbit coupling with the oxygen orbitals comparatively to the situation found in the "rapid"-giving species with its sulfur atom [104]. In fact, the "slow" signal-giving species is thought to share the same geometry with the species that gives rise to the "rapid type 1", with the sulfur being substituted by an oxygen and with the strongly coupled proton being assigned to that oxygen atom (Mo-OH_{strong} and Mo-SH_{strong}, in "slow" and "rapid" signals, respectively) [138]. Studies with model compounds further supported this interpretation [151]. As expected, the "slow" signal displays hyperfine structure due to two coupled oxygen atoms (A_{av} = 28 MHz) [145]. Hence, the structure of "slow" signal-giving species suggested by the EPR data is a **Mo^{5+}(=O_{axial})(-$OH_{equatorial}$)-$OH_{equatorial}$** complex (Fig. 3).

Signals Obtained From Inhibited and Inactive Forms of Xanthine Oxidase

The group of signals originated from inhibited XO molecules, in the sulfo and desulfo forms, are, obviously, not catalytically relevant, but the information obtained from their study is important to complete the structural picture of the enzyme molybdenum centre.

The "inhibited" signal is originated from XO molecules in the native, sulfo, form in the presence of formaldehyde (methanal) or methanol, which are known to inhibit the enzyme. It is a rhombic signal, with $g_1 < 2$ ($g_{1,2,3}$ = 1.991, 1.977, 1.951), and displays hyperfine structure due to a single non-exchangeable proton ($A_{1,2,3}$ = 12.3, 10.8, 15.3 MHz) (Fig. 4) [108, 122, 145, 158]. Generation of the "inhibited" signal with ^{2}H-labelled formaldehyde showed that the coupled proton is derived from the inhibitor and the use of ^{13}C-labelled formaldehyde ($A_{1,2,3}(^{13}C)$ = 52.5, 40.6, 40.6 MHz [108, 128]) suggested that the formaldehyde is coordinated to the molybdenum, probably in a bidentate mode [159] (Fig. 4). A more recent ENDOR study revealed that both protons of the formaldehyde moiety do interact with the molybdenum, but one of the couplings is too weak to be observed by standard CW EPR [160]. In accordance, it is proposed that the formate moiety becomes bound to the molybdenum, through a stable Mo^{5+} complex (Fig. 4), thus preventing the turnover and inhibiting the enzyme.

On the contrary, the "desulfo-inhibited" signal is originated from desulfo XO molecules. The "desulfo-inhibited" signal is obtained by developing, first, the "slow" signal in the presence of dithionite and, subsequently, adding ethylene-glycol (HO-CH_2-CH_2-OH) [161]. This denomination is derived from the XO form that gives rise to the signal ("desulfo") and its subsequent resistance to both oxidation and reduction ("inhibited"). It is also a rhombic signal, with $g_1 < 2$ ($g_{1,2,3}$ = 1.980, 1.973, 1.967); it does not display hyperfine structure due to protons, but it is coupled to two oxygen atoms (A_{av} = 5.5 MHz) [145, 161]. The "desulfo-inhibited" signal-giving species has been suggested to hold the ethylene-glycol moiety coordinated to the molybdenum, probably, also in a bidentate mode (Fig. 4).

The "arsenite" signal is obtained in the presence of arsenite, a XO inhibitor that binds particularly tightly to reduced XO [162, 163]. The signal is formed from reduced enzyme molecules either in the sulfo or desulfo form and displays strong anisotropic hyperfine and quadrupole coupling to the ^{75}As nucleus ($I = 3/2$ naturally present in 100%; $A_{1,2,3}(^{75}As)$ = −40, 128, −90 MHz and $P_{1,2,3}(^{75}As)$ = 27, −17,

−10 MHz) [162–165]. The signal is modified by xanthine, suggesting that the enzyme active site retains the ability to bind substrates in the presence of the inhibitor. Based on the known affinity of arsenite for thiols, it was suggested that arsenite binds to the reduced molybdenum center through its sulfo group, Mo-S-AsO_2^{2-} (Fig. 4).

The structure of the "arsenite" signal-giving species was also probed by XAS spectroscopy that suggested a non-linear Mo-S-As moiety (angle of ca 80°), with a Mo…As distance of 3.02 Å, and one of the arsenite oxygen atoms possibly bound to the molybdenum in the position of the equatorial Mo-OH (Fig. 4) [166]. The recent crystal structure of arsenite-inhibited XO clearly demonstrate the arsenite bidentate binding mode, involving the catalytically essential sulfo group and the equatorial labile oxygen atom [167], thus explaining the particularly tight binding of arsenite to reduced XO and the enzyme inhibition.

The "mercurial" signal is obtained when native, sulfo, XO is treated with *p*-chloro-mercuri-benzoate [168]. This thiol-modifying reagent reacts with the Mo-SH group of reduced XO, blocking it in a Mo-S-Hg-R complex, to yield an inactive XO form. This enzyme form gives rise to a rhombic signal ($g_{1,2,3}$ = 1.969, 1.958, 1.943) that does not display hyperfine structure due to coupled protons (Fig. 4) [168]. When the signal is generated in the presence of ^{199}Hg(I = 1/2)-labelled *p*-chloro-mercuri-benzoate a strong and anisotropic coupling is observable ($A_{1,2,3}$(^{199}Hg) = 443, 285, 272 MHz). The also strong and anisotropic coupling to ^{17}O ($A_{1,2,3}$(^{17}O) = 4.1, 8.2, 34 MHz) and ^{33}S ($A_{1,2,3}$(^{33}S) = 13.8, 7.4, 8.7 MHz), further supports the presence of a **Mo^{5+}(=O_{axial})(=$O_{equatorial}$)-$S_{equatorial}$-Hg(chloro-benzoate)** complex in the "mercurial" signal-giving species (Fig. 4).

Magnetic Interactions Within Xanthine Oxidase

XO holds two [2Fe-2S] centres and one FAD, besides the molybdenum centre. The two Fe/S centres, diamagnetic in the oxidised state ($[2Fe\text{-}2S]^{2+}$), can be spectroscopically discriminated upon reduction by one electron ($[2Fe\text{-}2S]^{+}$) to yield two S = 1/2 signals, called Fe/S I and II, with different g anisotropy and spin-relaxation rate [154, 169]. The Fe/S I signal, similar to the one spinach ferredoxin, displays lower g anisotropy ($g_{1,2,3}$ = 2.022, 1.932, 1.894) and, due to its slower spin-relaxation rate, can be observed at higher (liquid nitrogen) temperatures, comparatively to the Fe/S II signal ($g_{1,2,3}$ = 2.110, 1.991, 1.902, observed below 25 K, with unusual very broad linewidths). The one-electron reduced flavin gives rise to an isotropic S = 1/2 signal, with a linewidth of 1.94 mT characteristic of the neutral semiquinone radical, $FADH^{\bullet}$ [107, 154].

Because XO holds four redox-active centres, magnetic interactions between them are possible and expected. The magnetic interactions in XO have been studied since the 1970s, when the magnetic interaction between the Mo^{5+} and the Fe/S I was described for the first time: Ehrenberg, Anger and Bray observed that the "rapid" and "slow" signals are modified when the temperature is decreased [170]. In 1972, Lowe et al. [170] systematically studied the "slow" signal and concluded that its splitting is dependent on the presence and intensity of the signal originated by the Fe/S I, resulting from a spin-spin isotropic coupling of 1.1 mT (40 K). Yet, the predicted splitting of the signal from the reduced Fe/S I could not be observed. Later, Lowe and Bray [171], studying the "desulfo inhibited" signal, obtained a larger spin-spin coupling ($d_{1,2,3}$ = 2.2, 2.4, 2.6 mT) that allowed the corresponding splitting of the Fe/S centre signal to be also observed, thus providing positive identification of this Fe/S centre as the interacting species with the molybdenum. Similar magnetic interactions, whose intensity is dependent on the Mo^{5+} signal type (ranging from 0.7 mT for the "rapid" to 2.4 mT for the "desulfo inhibited"), were also described in turkey, *Veillonella alcalescens* [171] and rat liver XO [141], in *P. putida* quinoline 2-oxidoreductase [172], or in *D. gigas* AOR [173, 174].

Other magnetic interactions in XO were described by Rupp et al. [175] and Barber et al. [176], namely between Mo^{5+} and Fe/S I (identified as a 100-fold increase in the power necessary to saturate the Mo^{5+} signal in the presence of reduced Fe/S I, at 103K), Fe/S I and Fe/S II (2.5-fold increase in the power necessary to saturate the Fe/S I signal in the presence of reduced Fe/S II, at 20 K) and between FAD and Fe/S I and between FAD and Fe/S II (70-fold increase in the power necessary to saturate the $FADH^{\bullet}$ signal in the presence of reduced Fe/S centres, at 173 K). No interactions were described between Mo^{5+} and $FADH^{\bullet}$ or between Mo^{5+} and Fe/S II [176].

The magnetic interactions detected are in complete agreement with the structure of XO that shows that the four redox-active centres are aligned in an almost linear fashion, defining an intramolecular electron transfer pathway that delivers electrons from the molybdenum centre to the FAD, Mo → Fe/S I → Fe/S II → FAD [85]. Hence, it is the Fe/S centre that gives rise to the Fe/S I signal that is present in the nearer α-helical domain to the molybdenum center. Moreover, the distance between the Fe/S I centre and the molybdenum were estimated by Lowe and Bray to be 10–25 Å [171] and by Coffman and Buettner to be ≤14 Å [177], values that compare remarkably well with the distance determined in crystal structure, with Mo-Fe of 14.7 Å.

EPR Studies of Desulfovibrio Aldehyde Oxidoreductase

D. gigas AOR was first described by Moura et al. [178] and is believed to be an aldehyde scavenger, acting in a complex chain of electron transfer proteins that links the oxidation of aldehydes to the reduction of protons [179, 180]. *D. gigas* AOR was the first XO family member for which the crystal structure was determined [181–183]. It is structurally similar to mammalian XO, but it holds only two Fe/S centres (yet, when the AOR structure is represented with its putative physiological partner, one flavodoxin, it becomes clear that the structural homology with XO is preserved [184]).

D. gigas AOR gives rise to several Mo^{5+} signals, all similar to the ones of XO, as expected from two similar enzymes. After rapid reduction with benzaldehyde, salicylaldehyde, acetaldehyde or dithionite, AOR gives rise to a "rapid type 2" signal, with the characteristic two interacting protons ($g_{1,2,3}$ = 1.988, 1.970, 1.964, $A_{1,2,3}(^1H)$ = 32.0, 45.8, 34.6 and 31.7, 17.4, 25.6 MHz), as demonstrated by assays with ^2H-labelled water; the signal intensity indicated the presence of 10–30% of sulfo enzyme molecules [179, 180, 185, 186]. The formation of the "rapid type 1" signal can be achieved, e.g., with di-hydoxy-benzaldehyde reduction (Fig. 5) [18]. Extended reduction with dithionite (20 min) elicits the "slow" signal that arises from the remainder desulfo enzyme molecules present in the sample ($g_{1,2,3}$ = 1.971, 1968, 1.958, where only one coupled proton was detected, $A_{1,2,3}(^1H)$ = 46.4, 44.1, 39.5 MHz) [179]. So far, no signal arising from a AOR-substrate or AOR-product complex was reported (no "very rapid"-like signal was yet described).

Recent crystallographic results and kinetic assays showing that AOR is not irreversible inhibited by cyanide led Santos-Silva et al. to suggest that the active AOR holds a molybdenum centre with an oxo group in place of the XO characteristic sulfo group, (Mo=O instead of Mo=S) and that AOR did not require the sulfo group for activity [187]. Yet, the XO family members susceptibility to cyanide is highly variable. This is the case of the nicotinate dehydrogenase (described in the following section) or the purine hydroxylase [188, 189] that require extensive treatments to inhibit the enzyme. Hence, it can not be excluded that AOR is also an enzyme that is highly resistant to cyanide inactivation. This supposition is in agreement with the observation of "rapid" signals that are well documented to arise from MoO_2S cores, thus demonstrating that sulfo enzyme molecules are present in the *Desulfovibrio* AOR samples. However, the small percentage of the sulfo enzyme molecules could have not been observed by crystallography, being identifiable only by EPR: EPR spectroscopy allows the observation of sulfo and desulfo molecules (depending on the reductant and time of reduction; see discussion in XO, section "The "slow" signal"), while crystallography only shows the dominant fraction of desulfo enzyme molecules.

The inhibitory effect of arsenite was also evaluated in AOR. The "arsenite" signal obtained is analogous to the XO one ($g_{1,2,3}$ = 1.979, 1.972, 1.922, $A_{1,2,3}(^{75}As)$ = 60, 136, 120 MHz, $P_{1,2,3}(^{75}As)$ = 19, −10, −9 MHz) and the crystal structure of the AOR-arsenite complex showed the arsenite ion bound, in a monodentate mode, to the molybdenum atom at the equatorial position occupied by the catalytically labile Mo-OH, **Mo-O$_{\mathbf{eqautorial}}$-AsO$_2$(=O$_{\mathbf{equatorial}}$)** (the absence of the Mo=S group hampering the bidentate mode seen in XO) [190, 191].

EPR Studies of Nicotinate Dehydrogenase and Carbon Monoxide Dehydrogenase

Finally, the EPR signals of two other XO family members, whose molybdenum centres are different from the one of XO, should be here considered, namely the EPR signals of nicotinate dehydrogenase and carbon monoxide dehydrogenase.

Some prokaryotic enzymes of the XO family, in place of the terminal sulfo group, Mo=S, have, instead, a terminal seleno group, Mo=Se, that is also essential for the catalytic activity (Fig. 1b) [192–195]. This is the case of the *E. barkeri* nicotinate

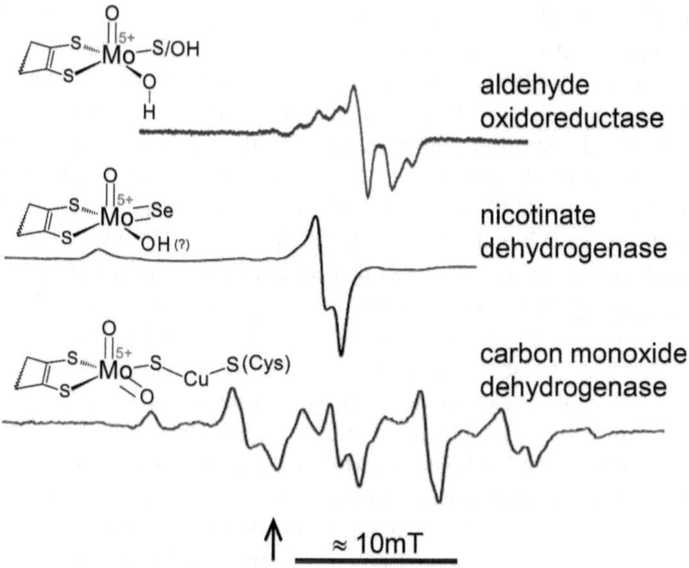

Fig. 5 Representative X-band EPR signals of aldehyde oxidoreductase, nicotinate dehydrogenase and carbon monoxide dehydrogenase and the respective structures proposed for each signal-giving species. The aldehyde oxidoreductase spectra is a mixture of "rapid" and "slow" signals and was obtained with dihydroxibenzaldehyde; adapted with permission from [18]. Nicotinate dehydrogenase "resting" (*as prepared*) signal; adapted with permission from [192], copyright (1994) National Academy of Sciences, U.S.A.. Carbon monoxide dehydrogenase signal obtained with carbon monoxide; adapted with permission from [200], copyright (2011) American Chemical Society. The *arrow* indicates the position of $g = 2.0037$

dehydrogenase, an enzyme that catalyses the hydroxylation of nicotinate to 6-hydroxy-nicotinate [192]. The reduced enzyme gives rise to an almost axial signal, with $g_1 \gg 2$ ($g_{1,2,3} = 2.067$, 1.982, 1.974 [192]), significantly higher than the g_1 value observed in the XO "very rapid" or "alloxanthine" signal (Fig. 5), as would be expected based on the higher covalency of the Mo-Se comparatively to a Mo-S. The signal is splitted when $^{77}Se(I = 1/2)$-enriched enzyme is used, showing that the selenium atom is located within magnetic contact to the molybdenum atom. Interestingly, the selenium-deficient enzyme (purified from selenium-deficient cells) gives rise to a Mo^{5+} signal similar to the "rapid type 1" of XO, suggesting that the molybdenum centres of nicotinate dehydrogenase and XO should be identical, apart from the replacement of Mo=S by Mo=Se, what was, in fact, recently confirmed when the crystallographic structure of the *E. barkeri* enzyme was determined [195].

Another "exception" to the characteristic XO molybdenum centre is provided by the Mo-S-Cu-S(Cys) binuclear centre of carbon monoxide dehydrogenase (Fig. 1b), an enzyme that catalyses the carbon monoxide oxidation to carbon dioxide [74–78]. This enzyme, upon reaction with carbon monoxide, gives rise to a unique Mo^{5+} signal, with $g_1 > 2$ ($g_{1,2,3} = 2.001$, 1.960, 1.955), that displays a strong anisotropic hyperfine coupling to one copper atom ($A_{1,2,3}(Cu) = 117$, 164, 132 MHz; $I = 3/2$ for both

^{63}Cu and ^{65}Cu, naturally present in ca 69 and 31%, respectively) (Fig. 5) [196–198]. A similar copper hyperfine interaction was observed in model complexes containing the Mo-S-Cu moiety [199]. When the signal is generated from carbon monoxide dehydrogenase molecules that were reconstituted with silver (after the sulfur and copper atoms had been removed by reaction with cyanide; the silver-substituted enzyme retains partially the activity [76, 200]), the expected splitted signal is observed ($A_{1,2,3}$(Ag) = 82.0, 78.9, 81.9 MHz; I = 1/2 for both ^{107}Ag and ^{109}Ag, naturally present in ca 52 and 48%, respectively) [200]. Moreover, the signal of CO-reduced enzyme in ^{1}H- or ^{2}H-labelled water displays no hyperfine coupling to protons, suggesting that this enzyme holds an equatorial deprotonated oxo group, Mo=O, contrary to XO with its labile Mo-OH group. Finally, the splitting of the Mo^{5+} signal, when ^{13}C-labelled carbon monoxide-reduced enzyme was used, demonstrated clearly that the carbon monoxide or dioxide is present in the signal-giving species of substrate-reduced enzyme [198]. Hence, the structure of the signal-giving species suggested by the EPR data is a **Mo^{5+}(=O_{axial})(=O)-S-Cu-S(cys)$_{equatorial}$** complex, with the carbon monoxide coordinated to the copper atom (thus, a paramagnetic enzyme-substrate complex analogue). This structure is supported (1) by the *O. carboxidovorans* enzyme crystallographic structure (the Mo-S-Cu-S(Cys) moiety) [75]; (2) by XAS studies that showed the presence of a two Mo=O groups and of an approximately linear Cu^{+} coordination [201]; (3) by resonance Raman spectroscopy, whose results are consistent with a $CuSMoO_2$ core [198, 201]; () and by a recent ENDOR study that, using ^{13}C-labelled carbon monoxide, showed an isotropic hyperfine coupling ($A_{av}(^{13}C)$ = 17.3 MHz) that could arise only from a Mo^{+5}/Cu^{+}-CO species [202].

Sulfite Oxidase Family

The Enzymes

The active site of SO family enzymes (in its oxidised form; Fig. 1b) is closely related with the one of XO family, but with the distinctive feature of having the protein, through a cysteine residue, directly coordinated to the molybdenum. In these enzymes, the molybdenum centre displays the same square-pyramidal geometry, with one apical oxo group (Mo=O), but with the equatorial plane formed by two sulfur atoms of one pyranopterin cofactor molecule, one oxo group (Mo=O) and the cysteine sulfur atom (Mo-S(Cys)) [4–11, 203, 204]. This family comprises the prototype vertebrate (most studied chicken and human) liver SO (Eq. (8)), plant SO, diverse prokaryotic sulfite dehydrogenases and eukaryotic assimilatory nitrate reductases (NaR; enzymes involved in nitrate assimilation in plants, algae and fungi[3]), as well as *Escherichia coli* YedY or mammalian mitochondrial amidoxime

[3] It should be noted that the eukaryotic assimilatory NaR is distinct from any type of prokaryotic NaR enzymes, which are classified as members of the dimethylsulfoxide reductase family (Section

reducing component (mARC; enzymes involved in the reduction (dehydroxylation) of S- and N-hydroxylated compounds) and the MOSC proteins homologues (involved in molybdenum centre sulfuration) [71–73, 203–208].

Structurally, the vertebrate SO is an homodimeric enzyme, with each monomer holding one *b*-type haem and one molybdenum centre [209]. Being a key enzyme in the catabolism of sulfur-containing amino acids and other compounds, vertebrate SO catalyses the oxidation of the toxic sulfite to sulfate, at the molybdenum centre (Eq. (9)), with the simultaneous reduction of cytochrome *c*, at the *b*-type haem [210]. Surprisingly, the crystal structure of vertebrate (chicken) SO showed that the molybdenum and haem centres are more than 30 Å apart [209]. Hence, during catalysis, a conformational alteration would have to take place, to bring the two centres into sufficiently close proximity as to allow rapid intramolecular electron transfer observed, (sulfite→)Mo→haem(→cytochrome *c*) [211–213].

The members of the SO family, in contrast to XO family enzymes, are thought to be proper oxo-transferases, catalysing the simple transfer of an oxygen atom to, or from, a lone electron pair of the substrate, as is clearly exemplified by the SO-catalysed sulfite oxidation to sulfate (Eq. (9)) and NaR-catalysed nitrate reduction to nitrite (Eq. (10)), respectively [4–11, 203, 204]. However, the recent identification of mammalian mARC and bacterial YedY, YcbX or YiiM, as well as several other MOSC proteins homologues (most of these not yet characterised), demonstrated that SO family enzymes are also involved in the reduction of S- and N-hydroxylated compounds and in sulfuration reactions.

$$SO_2^{2-} + H_2O + 2cyt.c\left(Fe^{3+}\right) \rightarrow OSO_2^{2-} + 2cyt.c\left(Fe^{2+}\right) + 2H^+ \tag{8}$$

$$SO_2^{2-} + H_2O \rightarrow OSO_2^{2-} + 2e^- + 2H^+ \tag{9}$$

$$ONO_2^{-} + 2e^- + 2H^+ \rightarrow NO_2^{-} + H_2O \tag{10}$$

The molecular mechanism of SO-catalysed sulfite oxidation is presently well understood (Fig. 6) [209, 214–222]: (1) catalysis is initiated at the oxidised molybdenum centre by nucleophilic attack of the sulfite lone-pair of electrons on the catalytically labile equatorial oxo group of the molybdenum (Mo^{6+}=O); this results in the formation of a covalent Mo^{4+}-O-SO_3 intermediate, where the molybdenum atom has become reduced by two electrons; (2) the presence of the "spectator" axial oxo group facilitates the subsequent cleavage of the Mo-O(substrate)$_{equatorial}$ bond (weakens it); product (sulfate) is then released to yield a Mo^{4+}-$OH_{(2)}$ core (the oxidation half-reaction); (3) finally, the two electrons transferred from the substrate to the molybdenum are intramolecularly transferred, one at a time, to the haem, where cytochrome *c* (the physiological partner) will be reduced, and the initial Mo^{6+}=O core is regenerated (reduction half-reaction).

"Dimethylsulfoxide Reductase Family").

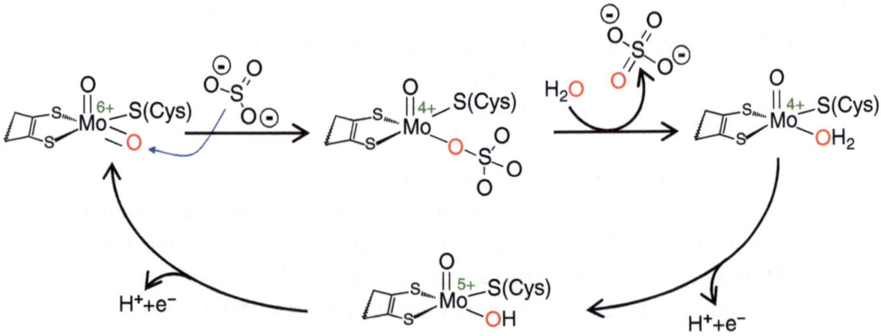

Fig. 6 Mechanism of sulfite oxidase-catalysed reaction. See text for details

In SO enzymes with one one-electron redox center (one haem in the case of chicken and human SO), the formation of a transient Mo^{5+} species is mandatory: the two electrons of Mo^{4+} must be, one at a time, transferred to the haem; after the haem being reduced by the first electron, it must wait for the physiological oxidising partner to be re-oxidised; only then, is able to receive the second electron. Hence, as in the XO family, the intramolecular electron transfer (Mo → haem, in this case) is an integral aspect of catalysis in the enzymes with one one-electron redox center. On the contrary, in the plant SO, that is devoid of additional redox centres, the reduction half-reaction is also carried out at molybdenum centre (in this case, the reduction of dioxygen) [223]. Also as seen in XO family, water is the ultimate oxygen atom donor/acceptor (Eqs. (9) and (10)) and the molybdenum centre mediates the oxygen atom transfer ($Mo^{6+}{=}O_{equatorial}$ is the direct oxygen donor and Mo^{4+} the oxygen acceptor).

EPR Studies of Vertebrate Sulfite Oxidase

Chicken SO is the best studied member of this family, with the human enzyme following closely, and this chapter will focus only the vertebrate enzymes (Fig. 7).

SO was the fifth enzyme discovered to contain molybdenum (after XO, AO, NaR and nitrogenase). The presence of molybdenum in bovine liver SO was revealed by Cohen, Fridovich and Rajagopalan, in 1971 [224], using EPR spectroscopy. Yet, comparatively to XO, the knowledge about SO advanced more slowly with relevant data being added only in the last decade.

Three Mo^{5+} EPR signals were early identified [224–230]. Generated by SO reduction with sulfite under low pH and high pH conditions and in the presence of phosphate, an inhibitor of the sulfite oxidising activity, they were designated "low pH" signal, "high pH" signal and "phosphate-inhibited" signal, respectively (Fig. 7). The "sulfite" signal was described by Bray et al. in 1982 [231], who pointed out its similarity with the "phosphate-inhibited" signal—both of which are now considered to arise from a family of similar signal-giving species, holding the anion coordinated via an oxygen atom to the molybdenum.

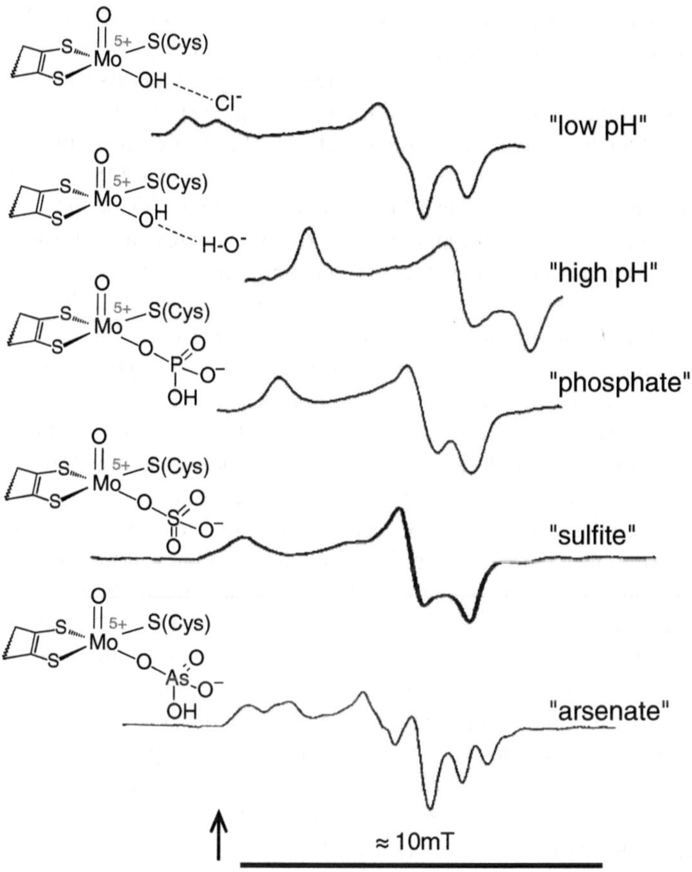

Fig. 7 Representative X-band EPR spectra of sulfite oxidase Mo^{5+} "low pH", "high pH", "phosphate", "sulfite" and "arsenate" signals and the respective structures proposed for each signal-giving species. "Low pH" signal obtained from sulfite-reduced SO at pH 6.5; adapted with permission from [230]. "High pH" signal obtained from sulfite-reduced SO at pH 10; adapted with permission from [230]. "Phosphate" signal obtained from sulfite-reduced SO, in the presence of phosphate buffer at pH 7.3; adapted with permission from [230]. "Sulfite" signal obtained from sulfite-reduced SO, at pH ≈ 6.4; adapted with permission from [231]. "Arsenate" signal obtained from sulfite-reduced SO, in the presence of arsenate at pH 6.5; adapted with permission from [260], copyright (1998) American Chemical Society. The *arrow* indicates the position of $g = 2.0037$

The "Low pH" Signal

The "low pH" signal, originated at moderately low pH values, is a rhombic signal, with $g_1 > 2$ ($g_{1,2,3}$ = 2.004, 1.972, 1.966), that displays strong coupling to one exchangeable proton ($A_{1,2,3}$ = 23.8, 22.1, 35.8 MHz) (Fig. 7) [229, 230]. More recently, it was also identified the coupling arising from the non-exchangeable proton of the Cα of the cysteine residue coordinated to the molybdenum atom ($A_{1,2,3} \approx$ 4, 6, 4 MHz) [232]. In the presence of ^{17}O-labelled water, the "low pH" signal exhibits strong coupling to a solvent-exchangeable oxygen [124], which, based on resonance Raman spectroscopic data, was assigned to the equatorial Mo-OH group [233]. A weak coupling to a solvent-exchangeable oxygen was also described in the human SO mutant where the key arginine 160 residue was replaced by a glutamine (Arg160Gln), in the presence of ^{17}O-labelled water [234]. Recently, this oxygen was attributed to the sulfite moiety bound to the molybdenum (Mo-OSO_3^-) [235, 236].

The interaction of chloride, an inhibitor of the vertebrate SO, on the signal-giving species was studied in the pH range of 6.2–9.6 [237]. Interestingly, it was found that the interconversion between the "low pH" and "high pH" signal is influenced by the chloride concentration, with higher concentrations favouring the "low pH" signal (that is, increasing the apparent conversion pK_a by 1 pH unit). Initially, it was proposed that chloride would be directly coordinated to the molybdenum [237] in an equatorial position [238, 239]. However, a more recent analysis of the ^{35}Cl and ^{37}Cl quadrupole interaction showed that, while the hyperfine coupling is within the range expected for coordinated chloride, the nuclear quadrupole coupling observed in SO is very close to that expected for free chloride [240]. Hence, supported by DFT calculations (that showed that the quadrupole coupling, contrary to the hyperfine coupling, is very sensitive to the centre structure), it was concluded that chloride is bound in the substrate binding pocket, near the molybdenum, but not directly bound to the molybdenum atom itself [240]. This interpretation is in agreement with the crystallographic structure of the chicken enzyme, that shows one chloride ion in the substrate binding pocket [241], and further confirmed by a recent XAS spectroscopic study with bromide and iodide that suggested a halide-molybdenum distance of 5 Å [242].

If an increase in chloride concentration favours the "low pH" signal, its depletion results in a new signal, with different g values and $g_1 < 2$ ($g_{1,2,3}$ = 1.999, 1.972, 1.963) and without the strong proton coupling [243], similar to that characteristic of the "blocked" form of SO [234, 244]. The "blocked" signal-giving species holds the sulfite moiety bound to the equatorial catalytic labile oxo group of the molybdenum (Mo-$O_{equatorial}$-SO_3^-) in a trapped complex, as shown by pulsed EPR spectroscopy with ^{17}O and ^{33}S labelling [234, 243–245]. Noteworthy, the subsequent addition of chloride causes the disappearance of the "blocked" signal, release of sulfate and the formation of the characteristic "low pH" signal [243]. It was, thus, proposed that chloride favours the cleavage of the Mo-OSO_3^- bond to yield sulfate and a Mo-OH moiety, which would give rise to the "low pH" signal with its coupled proton.

Noteworthy, the human SO Arg160Gln mutant remains in the "blocked" form [234, 246], even upon addition of chloride [243], holding sulfite/sulfate coordinated to the molybdenum (Mo-O-SO_3^-) [234]. This suggests that the product release in this mutant is very slow or hampered, which, together with its striking decreased intramolecular electron transfer rate, seems to be the base of the lethality of this mutation [234, 241, 243].

The mutation of the cysteine that coordinates the molybdenum to a serine residue, in recombinant rat SO (Cys207Ser), was also studied [247]. The oxidised molybdenum centre of the mutant, which displays a 2000-fold decreased specific activity, is a MoO_3 core that, unexpectedly, does not have the serine residue coordinated to it, as revealed by XAS spectroscopy [248] and further confirmed by the crystal structure and XAS of the mutant chicken SO [249]. Upon reduction, the serine becomes coordinated to the molybdenum, to give a Mo=O(-OH)-O(Ser) core, which gives rise to another EPR signal, with different g values and $g_1 < 2$ ($g_{1,2,3}$ = 1.979, 1.965, 1.955) and with a weaker proton coupling ($A_{1,2,3}$ = 6, 12, 6 MHz), comparatively to the wild-type SO [250].

On the contrary, the replacement of the molybdenum-coordinating cysteine by a selenocysteine residue resulted in a signal with an higher g_1 value ($g_{1,2,3}$ = 2.022, 1.975, 1.964), consistent with the higher covalency of the Mo-Se comparatively to a Mo-S (as describe above for the comparison between XO and nicotinate dehydrogenase; section "EPR Studies of Nicotinate Dehydrogenase and Carbon Monoxide Dehydrogenase") and with the known sensitivity of g_1 to modifications of equatorial plane of the molybdenum centre [251]. As expected, the new signal is also coupled to one proton ($A_{2,3}$ = 20, 35 MHz) [251].

The "High pH" Signal

The "high pH" signal, originated at higher pH values, is a rhombic signal, with $g_1 <$ 2 ($g_{1,2,3}$ = 1.987, 1.964, 1.953), that, contrary to the "low pH" signal, was described as not having hyperfine structure due to coupled protons (Fig. 7) [229, 230]. However, a later ENDOR study demonstrated the presence of a solvent-exchangeable, strongly and anisotropically coupled proton in this signal, which was also assigned to the equatorial Mo-OH group [252–254]. Different orientations of the proton in the non-linear Mo-O-H group relatively to the molybdenum have been evoked to explain its different coupling in the "low pH" and "high pH" group, with the hydrogen located "in" the ground-state orbital of the molybdenum and "out" of that orbital, respectively. The coupling arising from the non-exchangeable proton of the Cα of the cysteine residue coordinated to the molybdenum atom was also identified in the "high pH" signal [232].

The "high pH" signal displays, as the "low pH" signal, strong coupling to a solvent-exchangeable oxygen, when in the presence of ^{17}O-labelled water) [124], which was assigned to the equatorial Mo-OH group [233]. In addition, the "high pH" signal shows weak coupling to a second oxygen atom ($A_{av}(^{17}O)$ = 6.4 MHz, $P_{av}(^{17}O)$ = 1.5 MHz) [255], which was ascribed to the axial Mo=O group by comparison with the interaction observed in model complexes [147, 256].

Contrary to the "low pH", the depletion of chloride has no effect on the formation of the "high pH" signal-giving species [243] and the Arg160Gln mutation does not either change this signal [246]. Also the replacement of the molybdenum-coordinating cysteine by a selenocysteine causes almost no change in the "high pH" signal, with the most prominent effect being a modest increase in the g_2 value [251]. The small effect of having a selenocysteine at high pH (comparatively to the marked effect observed at low pH) is consistent with the lower covalency of the Mo-SeH bond at high pH (where increased ligand covalency results in higher g_1 values) [71, 257].

The Family of "Oxo-Anion" Signals

Similar signals can be obtained with some oxo-anions, at low pH and high anion concentrations, namely phosphate, arsenate and sulfite.

The so called "phosphate-inhibited" signal, originated in the presence of phosphate, an inhibitor of the sulfite oxidising activity, is a rhombic signal, with $g_1 < 2$ ($g_{1,2,3} = 1.992, 1.969, 1.974$), that does not display hyperfine structure due to coupled exchangeable protons (Fig. 7) [229, 230, 258, 259]. As noted in 1982 by Bray et al. [231], sulfite gives rise to a similar rhombic signal (but with g_1 close to 2), with no observable proton coupling (Fig. 7) [231, 243, 245]. In addition, also arsenate was shown to elicit an analogous signal (Fig. 7) [260].

The signal obtained with sulfite in the absence of chloride was later designated as "blocked" and, as already mentioned above (section "The "low pH" signal"), described to arise from a Mo-OSO_3^- core. The phosphate and arsenate complexes are thought to arise from similar complexes. Studies of the ^{31}P [258, 259] and ^{75}As [260] interactions ($I = 1/2$ and $I = 3/2$, respectively, both naturally present in 100%) indicated that the anions are coordinated to molybdenum through one of their oxygen atoms (Fig. 7), with the nuclear electric quadrupole coupling of ^{75}As pointing to a nearly tetrahedral arsenate [260]. Hence, a family of similar signal-giving species, holding the oxo-anion coordinated via one oxygen atom to the molybdenum, can be defined (Fig. 7).

Dimethylsulfoxide Reductase Family

The Enzymes

The DMSOR family is the larger and more diverse of the three families, comprising enzymes with widely varying function, subunit composition and makeup of additional redox-active centres [11, 24, 261–265]. The enzymes from this family harbour the molybdenum atom coordinated by four sulfur atoms of two pyranopterin cofactor molecules, in a trigonal prismatic geometry. In the oxidised form, the coordination sphere is completed by oxygen and/or sulfur and/or selenium atoms in a diversity of arrangements (Fig. 1b); the molybdenum atom is most often directly coordinated by the polypeptide chain via aspartate, serine, cysteine or selenocysteine residue side chains (arsenite oxidase, having no coordination to the polypeptide chain, is one important exception with the molybdenum centre coordinated by an apical oxo group plus one oxo/hydroxyl group (Mo=O(-OH)), in a square-pyramidal coordination geometry [266–269]). The DMSOR family is constituted by only prokaryotic enzymes that play remarkably different functions, including DMSOR, three different types of NaR enzymes (dissimilatory membrane-bound enzymes, dissimilatory periplasmatic enzymes and assimilatory cytoplasmatic enzymes (see note 3)), arsenite oxidase, arsenate reductase, several different types of formate dehydrogenases (FDH), polysulfide reductase, acetylene hydratase and many other.

Matching that diverse functionality, these enzymes catalyse remarkably different reactions, namely (1) proper transfer of oxygen atom (e.g., DMSOR-catalyse DMSO reduction (Eq. (11)) or NaR-catalysed nitrate reduction (Eq. (10))), (2) transfer of hydrogen atom (reversible FDH-catalysed formate oxidation to carbon dioxide (Eq. (12)) or benzoyl-coA reductase-catalysed benzoyl-CoA reduction to cyclohexa-1,5-diene-1-carboxyl-CoA), (3) transfer of sulfur atom (e.g., polysulfide reductase-catalysed inorganic sulfur reduction to sulfide), (4) *simultaneous* oxidation and reduction (e.g., reductive dehydroxylation and concomitant oxidative hydroxylation catalysed by pyrogallol:phloroglucinol hydroxyltransferase) and (5) even hydration reactions (e.g., acetylene hydratase-catalysed hydration of acetylene to acetaldehyde, a non-redox reaction).

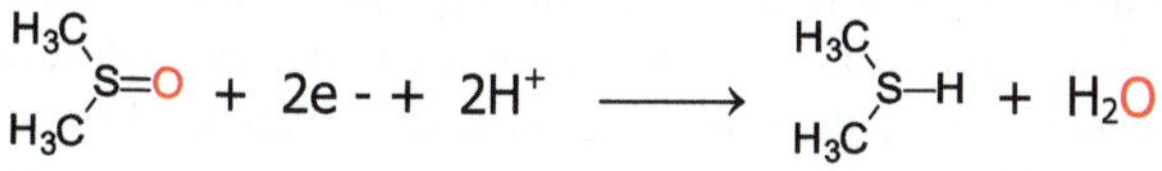

(11)

$$HCOO^- \rightarrow CO_2 + 2e^- + H^+ \quad (12)$$

To restrict the information presented to a manageable size, only two enzyme types from the DMSOR family will be described, namely the NaRs and FDHs.

Prokaryotes reduce nitrate to nitrite for dissimilatory and assimilatory processes and encode three distinct NaR enzymes [24, 270–278]: (a) membrane-bound cytoplasm-faced respiratory NaR, associated with the generation of a proton motive force across the cytoplasmatic membrane; (b) periplasmatic NaR, involved in the generation of a proton motive force or acting as an electron sink to eliminate excess of reducing equivalents; and (c) cytoplasmatic assimilatory NaR, involved in nitrogen assimilation. In spite of catalysing the same reaction (Eq. (10)), that occurs at the molybdenum centre, these enzymes have diverse subunit organisations and cofactor compositions and harbour different molybdenum centres. The three types of NaR enzymes have the molybdenum atom coordinated by the family characteristic two pyranopterin cofactor molecules (Fig. 1), but the remainder of the molybdenum coordination sphere is distinct in each enzyme type. In the respiratory membrane-bound NaR, the molybdenum atom is further coordinated by an aspartate residue in a bidentate fashion (that is, by the two oxygen atoms of its carboxylate), or alternatively by one terminal oxo group plus the aspartate residue coordinated in a monodentate mode (that is, by only one of its carboxylate oxygen atoms) [272, 279]. The two aspartate binding modes are thought to correspond to oxidised and reduced molybdenum centre. In the case of the periplasmatic NaR from *Desulfovibrio desulfuricans* or *Cupriavidus necator*, the molybdenum atom is coordinated instead by a cysteine sulfur atom plus one terminal sulfo group, forming a partial disulfide bond within each other [280–283]. Yet, the *E. coli* and *Rhodobacter sphaeroides* periplasmatic NaR, on their turn, complete the molybdenum coordination with the cysteine sulfur atom plus a terminal hydroxyl group [284, 285]. Finally, the cytoplasmatic assimilatory NaR, the least studied one, harbours its molybdenum atom coordinated by a cysteine residue [286].

The FDH enzymes are involved in different biochemical pathways, being also a structurally heterogeneous group of enzymes [24–26]. *E. coli*, e.g., expresses three FDHs: (a) a cytoplasmatic enzyme designated formate dehydrogenase H (FDH-H) [287] that is part of the formate-hydrogen lyase system and is involved in the oxidation of formate and generation of molecular hydrogen formation under fermentative (anaerobic) growth conditions [24, 288]; (b) a membrane-bound periplasm-facing FDH designated formate dehydrogenase N (FDH-N) that is part of the anaerobic nitrate-formate respiratory pathway (catalysed by a supermolecular formate:nitrate oxidoreductase system formed with the NaRGHI) involved with the generation of a proton motive force [264, 289–295]; and (c) a second membrane-bound periplasm-faced FDH, designated formate dehydrogenase O, that operates under aerobic conditions in another nitrate-formate respiratory pathway (this with the NaRZWV enzyme) [296–300]. In the case of FDHs, contrary to the above described NaRs, the molybdenum centre, where the reversible formate oxidation takes place (Eq. (12)), is more "uniform". Hence, these enzymes, in the oxidised form, have the molybdenum atom coordinated by two pyranopterin cofactor molecules and by a terminal sulfo group plus a conserved essential selenocysteine or cysteine residue (Mo=S(-Se(Cys)) or Mo=S(-S(Cys)); Fig. 1) [292, 301–309].

EPR Studies of Formate Dehydrogenases

The formate-reduced *E. coli* FDH-H gives rise to a nearly axial signal ($g_1 \gg g_2 \approx g_3$), with $g_1 = 2.094$ ($g_{2,3} = 2.001, 1.990$) that displays coupling to one solvent-exchangeable proton ($A_{1,2,3} = 7.5, 18.9, 20.9$ MHz), designated "2.094" signal (Fig. 8) [310, 311]. When the signal is generated from ^{77}Se-enriched enzyme a very strong and anisotropic interaction is observed ($A_{1,2,3}(^{77}\text{Se}) = 13.2, 75, 240$ MHz [312], values that are almost five times higher than the ones observed in Mo-Se model compounds [116, 312]). This interaction and the observation of the expected ^{95,97}Mo hyperfine coupling confirmed that the selenium atom of the selenocysteine is directly coordinated to the molybdenum and suggested that the unpaired electron is delocalised 17–27% over the selenium, once more supporting the covalency introduced by selenium in the Mo–Se bond [312].

Studies with ^{2}H-labelled formate (*E. coli* FDH-H) showed that the coupled solvent-exchangeable proton is originated from the substrate molecule [312]. It is worth mentioning that the proton coupling is absent in the beginning of reaction and it appears only in the following 30–300 s of reaction with formate [312]. The hyperfine interaction demonstrated that the proton acceptor is located within magnetic contact to the molybdenum center and photolysis assays showed that the selenium remains bound to the molybdenum and that it is not the proton acceptor (because in the photo-converted enzyme centre the interaction with ^{77}Se is not significantly affected, while the interaction with the proton disappears) [312]. Together, these results clearly demonstrate that the proton is transferred from formate Cα to the molybdenum center in the course of the catalytic turnover and, then, exchanged with the solvent. Based on these observations and on the crystallographic structure determined Boyington et al. [303]

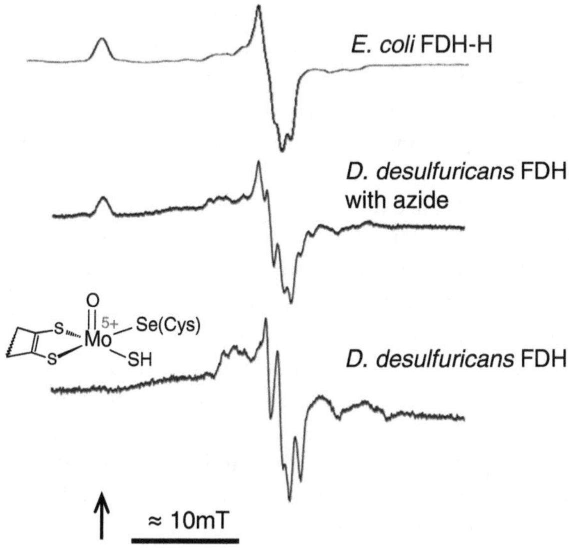

Fig. 8 Representative X-band EPR spectra of formate dehydrogenases. The *top spectrum* was obtained with formate-reduced *E. coli* FDH; adapted with permission from [312], copyright (1998) American Chemical Society. The *middle spectrum* was obtained from formate-reduced *D. desulfuricans* FDH in the presence of azide; adapted with permission from [307]. The *bottom spectrum* was obtained from formate-reduced *D. desulfuricans* FDH (without azide); adapted with permission from [307]. It is represented the signal-giving species structure proposed in [314, 315]. The *arrow* indicates the position of $g = 2.094$

and XAS data [313], Khangulov et al. [312] suggested that the *E. coli* FDH-H signal-giving species would have a Mo^{+5}-Se(Cys) core, with the coupled formate-derived proton residing on an amino acid residue located very close to the molybdenum (His_{141}). However, the subsequent identification of a sulfo group in the coordination sphere of the *E. coli* FDH-H molybdenum by Raaijmakers and Romão [306] (also confirmed by the identification of an *E. coli* sulfurtransferase that would insert the sulfur atom into the FDH molybdenum centre [308]), provided a more reasonable acceptor for the coupled formate-derived hydrogen: the Mo=S group that would be converted into a Mo-SH, thus, a direct molybdenum ligand being the proton acceptor [314, 315]. Hence, the suggested structure of the signal-giving species is a **Mo^{5+}-SH(-Se(Cys))** complex [314, 315] (Fig. 8).

The crystallographically proposed dissociation of the selenocysteine from the molybdenum coordination sphere by Raaijmakers and Romão [306] does not find support in none of the so far described EPR studies (nor in the *E. coli* enzyme EXAFS study at both the Mo and Se K-edges that indicated four or five Mo-S at 2.36 Å, plus one Mo-Se at 2.62 Å [313], or in the crystallographic structures of the *E. coli* FDH-N [292], *D. gigas* FDH [316]). It is possible that either the Mo^{5+} signal-giving species bears no relation to the species crystallographically observed or the crystallisation/irradiation induced some artefacts.

Subsequent studies by Rivas et al. [307] with the *D. desulfuricans* FDH demonstrated that the "2.094" signal only arises in the presence of azide, a strong FDH inhibitor (which was employed as a protective reagent during the *E. coli* FDH-H purifications). In fact, when the *D. desulfuricans* FDH is incubated with azide and reduced with formate or dithionite, the Mo^{5+} signals obtained showed almost identical g values and hyperfine structure to that described for the *E. coli* enzyme (Fig. 8) [307]. However, in the absence of azide, the formate or dithionite-reduced *D. desulfuricans* FDH gives rise to a rhombic signal with small anisotropy, with $g_{1,2,3}$ values of 2.012, 1.996, 1.985 [307]. The *D. desulfuricans* signal shows the expected hyperfine interaction with one solvent-exchangeable proton ($A_{1,2,3}$ = 23.1, 29.9, 27.8MHz) and an additional interaction with a second not solvent-exchangeable proton (A_1 = 35.1 MHz, $A_{2,3}$ not detectable) [307]. This later interaction arises from the selenocysteine Cβ hydrogen atoms [307], while the former, as discussed above for the *E. coli* enzyme, is assigned to the molybdenum sulfo group [314, 315].

With the *D. desulfuricans* enzyme, the studies with ^{2}H-labelled formate were inconclusive, because the $^2H/^1H$ exchange with the solvent is faster than the timescale of the "freezing technique" used. However, in the presence of the inhibitor azide, the reaction is decelerated and it was clearly shown that the formate Cα hydrogen is transferred, within 5 s, to a ligand of the molybdenum atom ($A_{2,3}$ = 21.0, 21.1 MHz, A_1 not determined) and, then, exchanged rapidly with the solvent [307]—thus, the strongly coupled proton is substrate-derived and, then, solvent-exchangeable. In this way, the EPR-supported structure for the signal-giving species is, once more, a **Mo^{5+}-SH(-Se(Cys))** complex [314, 315] (Fig. 8). Although the crystallographic structure of the *D. desulfuricans* FDH was not yet determined, the XAS results obtained suggested that both the oxidised and reduced molybdenum center is hexa-coordinated, with the selenocysteine and sulfo group bound to the molybdenum atom [317], thus supporting that signal-giving species structure.

Among other FDH enzymes, the NAD-dependent FDH from *Methylosinus trichosporium* and *Ralstonia eutropha* were also probed by EPR spectroscopy. The *M. trichosporium* enzyme displays a similar rhombic Mo^{5+} signal with small anisotropy ($g_{1,2,3}$ = 2.005, 1.091, 1.984) and well-resolved hyperfine structure due to ^{95,97}Mo [318]. This is also the case of the *R. eutropha* FDH Mo^{5+} signal ($g_{1,2,3}$ = 2.009, 2.001, 1.992 and $A_{1,2,3}(^{95,97}Mo)$ = 138, 82, 45 MHz) that displays strong isotropic coupling to one proton ($A_{1,2,3}(^1H)$ = 18, 21, 18 MHz) [314]. When the signal is generated in ^{2}H-labelled water, the proton hyperfine structure disappears, demonstrating the solvent-exchangeability of the proton. As observed in the *E. coli* and *D. desulfuricans* enzymes, studies with ^{2}H-labelled formate (in ^{1}H-water) confirmed that the solvent-exchangeable coupled proton is originated from the substrate molecule, with the coupling being absent in the beginning of the reaction and growing in over the course of 1 min of reaction, to yield a signal indistinguishable from that seen in the dithionite-reduced enzyme. Hence, additional proof that the proton is transferred from formate Cα to the molybdenum center in the course of the reaction and, then, exchanged with the solvent. Therefore, there is a general consensus that the structure of the signal-giving species in FDHs is a **Mo^{5+}-SH(-Se(Cys))** complex [314, 315]. This consensus gives confidence to the recent proposal of the FDH reaction mechanism involving the hydride transfer from formate to the molybdenum sulfo group, Mo=S, to yield carbon dioxide and a Mo-SH core [314, 315].

Besides the studies above described, the EPR spectroscopy was also successfully used to probe the magnetic interactions between the redox-active centres of several FDH enzymes, as is the case of the enzymes from *M. trichosporium* [318], *Desulfovibrio* sp. [319] and *R. eutropha* [314]. Moreover, and most relevant, EPR spectroscopy was essential to demonstrate the incorporation of either molybdenum or tungsten in *D. alaskensis* FDH, one of the few examples of an enzyme that can incorporate either metal atoms and retain activity [319]. Metal quantification assays indicated that the *D. alaskensis* FDH is a mixture of two forms, one containing one molybdenum atom and the other one tungsten atom. Two EPR signals, typical of Mo^{5+} and W^{5+}, were observed. The Mo^{5+} signal is rhombic, with small anisotropy and $g_1 < 2$ ($g_{1,2,3} = 1.971, 1.968, 1.959$) and displays hyperfine structure due to a coupled proton ($A_{1,2,3} = 44.2, 44.1, 43.9$ MHz). The W^{5+} signal is also rhombic, but with lower and more anisotropic g values ($g_{1,2,3} = 1.955, 1.933, 1.916$), as would be expected from the much stronger spin-orbit coupling in tungsten compared to molybdenum. Also because the CW EPR linewidths tend to be broader in tungsten compounds compared to molybdenum, no hyperfine structure is visible in the *D. alaskensis* FDH W^{5+} signal. The molybdenum substitution by tungsten was also studied in DMSOR (the substitution was forced; contrary to *D. alaskensis* FDH, the *Rhodobacter capsulatus* DMSOR does not incorporate tungsten naturally) [320]. Two different signals were observed, both rhombic, with higher g anisotropy, but, while one signal displays no visible hyperfine coupling to protons ($g_{1,2,3} = 1.958, 1.928, 1.860$, $A_{1,2,3}(^{183}W) = 104, 108, 110$ MHz), the other shows clear hyperfine coupling to one proton ($g_{1,2,3} = 1.960, 1.927, 1.888$, $A_{1,2,3}(^{183}W) = 110, 108, 106$ MHz, $A_{1,2,3}(^{1}H) = 35.7, 37.8, 52.9$ MHz).

EPR Studies of Nitrate Reductases

Periplasmatic Nitrate Reductases Signals

Periplasmatic NaRs from different sources were investigated by EPR spectroscopy, including from *D. desulfuricans* [274, 321, 322], *Paracoccus pantothrophus* [323–326], *R. sphaeroides* [285, 327, 328], *E. coli* [284] and others [329, 330]. The Mo^{5+} of these enzymes can give rise to several EPR signals, designated "very high *g*", "high *g*", "high *g*-nitrate", "high *g*-turnover", "low *g*-unsplit" and "low *g*-split".

The "very high *g*" is a rhombic signal, with $g_1 > 2$ that is coupled to one proton ($g_{1,2,3} = 2.022, 1.999, 1.994$, $A_{1,2,3} = 20.9, 20.7, 18.4$ MHz, for the *P. pantothrophus* enzyme); it is thought to arise from an inactive form of the enzyme (Fig. 9) [323, 324, 328]. The "low *g*" signals are also rhombic, but with $g_1 < 2$ ($g_{1,2,3} = 1.997, 1.962, 1.959$ and $g_{1,2,3} = 1.996, 1.969, 1.961$, for the "low *g*-unsplit" and "low *g*-split" signals of the *P. pantothrophus* enzyme) and the "low *g*-split", as its name

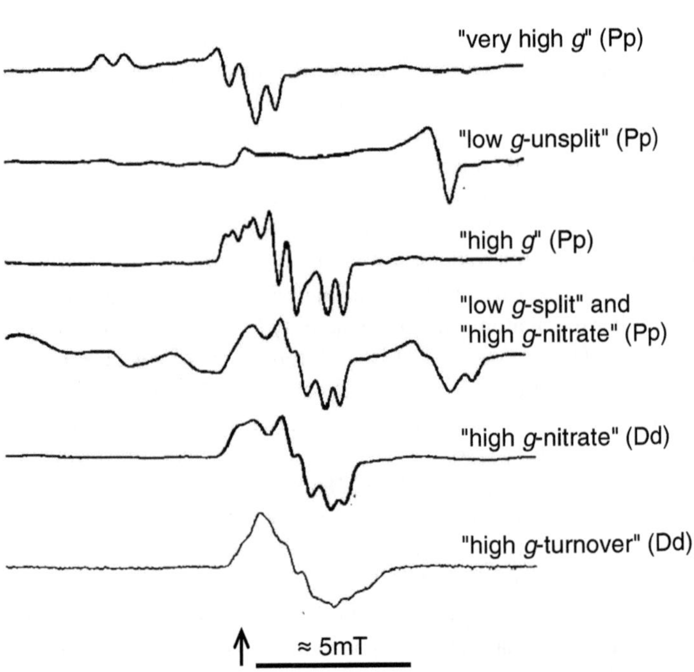

Fig. 9 Representative X-band EPR spectra of periplasmatic nitrate reductases. From *top* to *bottom*: "Very high *g*" signal obtained from cyanide-treated dithionite-reduced *P. pantothrophus* enzyme; adapted with permission from [324], copyright (1999) American Chemical Society. "Low *g*-unsplit" signal obtained from *P. pantothrophus* enzyme by prolonged incubation with dithionite; adapted with permission from [324], copyright (1999) American Chemical Society. "High *g*" signal was obtained from as-prepared, resting *P. pantothrophus* enzyme; adapted with permission from [324], copyright (1999) American Chemical Society. The fourth spectrum shows the "high *g*-nitrate" and "low *g*-split" signals obtained from dithionite-reduced *P. pantothrophus* enzyme that was reacted with nitrate; adapted with permission from [324], copyright (1999) American Chemical Society. "High *g*-nitrate" signal obtained from dithionite-reduced *D. desulfuricans* enzyme reacted with nitrate; adapted with permission from [321]. "High *g*-turnover" signal obtained from methyl-viologen-reduced *D. desulfuricans* enzyme reacted with nitrate; adapted with permission from [321]. The *arrow* indicates the position of $g = 2$

indicates, displays coupling to one proton ($A_{1,2,3}$ = 36.3, 37.5, 42.0 MHz) (Fig. 9) [323, 324, 328]. The "low *g*" signal is similar to the "rapid type 1" signal of XO and the signal-giving species is thought to hold the molybdenum atom coordinated by only one pyranopterin cofactor molecule. Furthermore, XAS results suggested the presence of at least one terminal oxo group [324]. Hence, the "low *g*-split" is also attributed to an inactive form of the enzyme.

The "high *g*" is a rhombic signal, with $g_1 < 2$ ($g_{1,2,3}$ = 1.999, 1.990, 1.981, for the *P. pantothrophus* enzyme) that displays weak coupling to two not solvent-exchangeable protons ($A_{1,2,3}$ = 17.9, 14.5, 13.9 MHz and A_1 = 8.4 MHz (*P. pantothrophus*)), which were assigned to the molybdenum-coordinated cysteine Cβ hydrogen atoms (Fig. 9) [323–326]. While the "high *g*" signal does not change in the presence of nitrate (as well as, of nitrite or azide), a different signal, the "high *g*-nitrate" signal, is obtained after the addition of nitrate to dithionite-reduced enzyme under turnover conditions, which has lower *g* anisotropy and one (*D. desulfuricans*) or two (*P. pantothrophus*) coupled protons ($g_{1,2,3}$ = 2.000, 1.990, 1.981, $A_{1,2,3}$ = 12.9, 13.9, 12.8 MHz, for *D. desulfuricans* [321, 322]; $g_{1,2,3}$ = 1.999, 1.989, 1.982, $A_{1,2,3}$ = 17.9, 12.0, 12.8 MHz and A_1 = 9.0 MHz, for *P. pantothrophus* [324, 325]) (Fig. 9). If methyl-viologen-reduced enzyme is used instead, the addition of nitrate elicits a similar signal, designated "high *g*-turnover", that displays hyperfine structure due to two equivalent solvent-exchangeable protons ($g_{1,2,3}$ = 1.999, 1.992, 1.982, $A_{1,2,3}$ = 16.2, 18.1, 15.3 MHz, for *D. desulfuricans* [321, 322]) (Fig. 9). Assays with common (^{14}N) and ^{15}N-labelled nitrate excluded the possibility that these signals arise from Mo-ONO_2 complexes, thus questioning the catalytic relevance of the signals [321, 322, 326], even though this conclusion is not consensual [328].

Presently, it is clear that the spectroscopic characterisation of the periplasmatic NaR enzymes is far from being concluded and the structure of the various signal-giving species is still obscure, as is their catalytic relevance (compared, e.g., with the wealth of information regarding the several XO signals (section "Xanthine Oxidase Family")). A recent theoretical study provided a rationalisation for the structure of some signal-giving species [331]. The "very high *g*" signal and "high *g*" signals family are suggested to arise from MoS_6 cores, with the "high *g*" species considered to be the ones in the more reduced state (that is, the ones with the proposed persulfide bond between the sulfo group, Mo-S, and the cysteine sulfur, Mo-S(Cys), [322] reduced) [331]. The differences between the signals may arise from different distortions of the molybdenum coordination sphere and/or the presence of different ions in the close proximity of the molybdenum [332]. The "low *g*" signals are suggested to arise from molybdenum centres with only one pyranopterin molecule, as suggested before (and described above), with the molybdenum coordination sphere completed as a **Mo=X(-OH)(-S(Cys))** core, where X is an oxygen or sulfur atom [331].

Assimilatory Nitrate Reductases Signals

The assimilatory NaRs are the least well characterised of these enzymes, but their Mo^{5+} signals are quite similar to the ones of the periplasmatic homologues, consequence of the presence of one molybdenum-coordinated cysteine in both enzymes types. The assimilatory enzyme from *Azotobacter vinelandii* enzyme is known for long to give rise to a "very high *g*" and "high *g*-nitrate" signals [333]. Also the cyanobacterium *Synechococcus* sp. PCC 7942 assimilatory enzyme elicits a "very high *g*" ($g_{1,2,3}$ = 2.023, 1.998, 1.993) and "high *g*" signals ($g_{1,2,3}$ = 1.997, 1.990, 1.982), which are thought to arise from inactive and functional enzyme forms, respectively [286].

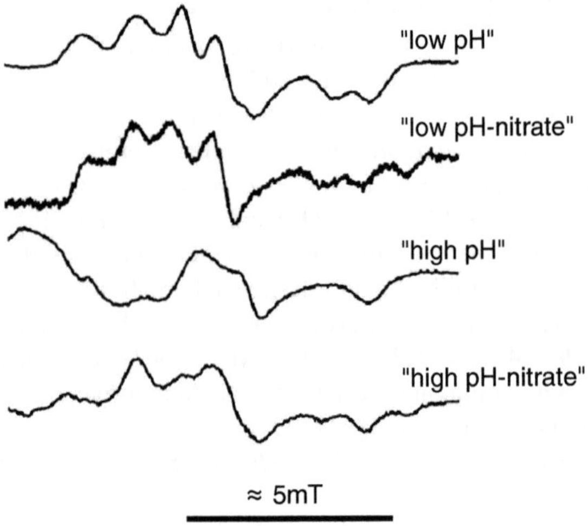

Fig. 10 Representative X-band EPR spectra of respiratory nitrate reductases from *D. desulfuricans*. "Low pH" signal obtained at pH 6.0. "Low pH – nitrate" signal obtained at pH 6.0 from dithionite-reduced enzyme treated with nitrate showing the "low pH – nitrate" signal plus a new rhombic signal. "High pH" signal obtained at pH 7.6. "High pH – nitrate" signal obtained at pH 6.0 from dithionite-reduced enzyme treated with nitrate showing the "low pH – nitrate" signal plus a new rhombic signal. All spectra were adapted with permission from [342]. The *arrow* indicates the position of $g = 2$

Respiratory Nitrate Reductases Signals

The Mo^{5+} of respiratory NaRs also give rise to several EPR signals, but distinct from the ones of the perplasmatic and assimilatory enzymes [334–342]. Two are pH-dependent and display hyperfine coupling to a solvent-exchangeable proton. The "low pH" is a rhombic signal, with $g_1 > 2$ ($g_{1,2,3} = 2.001, 1.986, 1.964$, $A_{1,2,3}(H) = 31.7, 23.6, 24.7$ MHz), while the "high pH" has a $g_1 < 2$ and a weaker hyperfine interaction ($g_{1,2,3} = 1.988, 1.981, 1.962$, $A_{1,2,3}(H) = 10.6, 8.9, 9.1$ MHz, for the *E. coli* enzyme) [338]. Similar signals have been since reported for the enzyme from *M. hydrocarbonoclasticus* enzyme (Fig. 10) [342], *Pseudomonas aeruginosa* [337] and others. Of note, the *P. pantothrophus* enzyme exhibits also "low pH" and "high pH" signals lacking resolved hyperfine coupling ($g_{1,2,3} = 2.007, 1.987, 1.970$ and $1.990, 1.989, 1.967$, respectively), which have been interpreted as arising form a different substrate binding mode [259, 338, 343–345]. The difference in the extent of coupling as a function of pH may arise from different orientations of the proton, in a non-linear Mo-O-H group, relatively to the molybdenum atom (as argued above for SO (section "Sulfite Oxidase Family")), or may be the result of different nitrate binding modes, as suggested for the *M. hydrocarbonoclasticus* enzyme [342]. Regarding the Mo-OH group, the "low pH" signals exhibits no ^{17}O hyperfine coupling in assays with ^{17}O-labelled water or ^{17}O-labelled nitrate, thus questioning the presence of a Mo-OH group in that signal-giving species (although it is possible that a very weak coupling could have not been observed in standard CW EPR assays); the "high pH" signal, in assays with ^{17}O-labelled water, on the other hand, displays a weak hyperfine coupling ($A_{1,2,3}(^{17}O)$ = 5.5, 8.8, 5.5 MHz), suggesting a Mo=O group [346]. The catalytic significance of the "low pH" signal has been defended since the late 1970s [336, 340], but a later study [336, 340] suggested that both signal-giving species would be involved in the enzyme turnover.

The effect of different anions in the "low pH" signal has also been studied. Fluoride ($I = 1/2$, naturally present in 100%) elicits a well defined hyperfine interaction ($A_{1,2,3}(^{19}F)$ = 30.6, 15.3, 13.2 MHz (*E. coli*)), suggesting that the halide is bound very near the molybdenum atom (instead of being directly coordinated to the metal, as argued above for SO (section "Sulfite Oxidase Family")) [338]. Addition of nitrate or nitrite to the *E. coli* enzyme give rise to different signals, with $g_1 > 2$, all showing hyperfine coupling to protons ($A_{av}(^{1}H)$ = 34–36 MHz) [338]. Similar signals were obtained upon nitrate addition to dithionite-reduced *M. hydrocarbonoclasticus* enzyme ($g_{1,2,3}$ = 2.002, 1.987, 1968, $A_{1,2,3}(^{1}H)$ = 39.2, 30.6, 30.3 MHz), but in a pH-independent manner and with the simultaneous observation of a new rhombic signal that displays no hyperfine structure ($g_{1,2,3}$ = 1.996, 1.982, 1979) (Fig. 10) [342].

Concluding Remarks

EPR spectroscopy has contributed enormously and significantly to our present knowledge about molybdenum-containing enzymes. This is particularly true for mammalian XO, to which the huge number of EPR studies (together with other spectroscopic methodologies), were decisive to draw a picture of the molybdenum centre of XO family enzymes before the first crystal structure of a XO family member was available in 1995 (*D. gigas* AOR). The EPR spectroscopy contribution to validate the hydroxylation reaction mechanism was also decisive, with the demonstration of hydrogen transfer from the substrate to the molybdenum sulfo group, Mo=S, and transfer of the water oxygen to the catalytically labile equatorial oxo group, Mo-OH. The contribution of EPR (and other spectroscopies) has also been extremely significant to our understanding of the SO molybdenum centre and reaction mechanism, namely the formation of "blocked" species, the chloride role or the mutant-based lethality. Yet, several questions remain open, including what drives the "low pH"/"high pH" transition (which residues) and the exact structures of those signal-giving species. Numerous questions remain also open regarding the FDH and NaR enzymes, for which it is clear that the spectroscopic characterisation is far from being concluded and the structure of the various signal-giving species is still obscure, as is their catalytic relevance. Nonetheless, the knowledge and experience gathered with the "older" enzymes and diverse model compounds will be, for sure, decisive to guide the future spectroscopic characterisation of the enzymes that only now are starting to be thoroughly studied.

Besides the mechanistic and structural studies focused in this chapter, the EPR spectroscopy has also been extensively used to monitor redox titrations. The knowledge of the reduction potentials is essential to understand the intramolecular electron-transfer processes within enzymes with several redox-active centres or between enzymes and their redox partners. In these studies, an EPR signal is taken as a measure of a determined oxidation state/enzyme species; e.g., the XO "rapid" and "slow" signals were used as indicators to assist in the determination of the reduction potential of the sulfo and desulfo molybdenum centres. Also the study of magnetic interactions is decisive to establish/confirm the "intramolecular wire of redox-active centres" that is responsible for the intramolecular electron-transfer, as was illustrated with the XO "intramolecular wire" that delivers the electrons from the molybdenum centre to the FAD.

Hence, we hope that we have managed to transmit that CW EPR is still a very valuable spectroscopic tool for the study of molybdenum-containing enzymes (and other metalloenzymes), although advanced EPR-related methods would be, for certainly, decisive.

Acknowledgements This work was supported by the Unidade de Ciências Biomoleculares Aplicadas—UCIBIO which is financed by national funds from FCT/MEC (UID/Multi/04378/2013) and co-financed by the ERDF under the PT2020 Partnership Agreement (POCI-01-0145-FEDER-007728). LBM thanks Fundacão para a Ciência e a Tecnologia, MEC, for a fellowship grant (SFRH/BPD/111404/2015, which is financed by national funds and co-financed by FSE).

References

1. Palmer G (1985) Biochem Soc Trans 13:548
2. Zhang Y, Gladyshev VN (2008) J Mol Biol 379:881
3. Zhang Y, Rump S, Gladyshev VN (2011) Coord Chem Rev 255:1206
4. Hille R (1996) Chem Rev 96:2757
5. Hille R (2002) Trends Biochem Sci 27:360
6. Schwarz G, Mendel RR, Ribbe MW (2009) Nature 460:839
7. Hille R, Mendel R (2011) Coord Chem Rev 255:991
8. Mendel R, Kruse T (2012) Biochim Biophys Acta 1823:1568
9. Hille R (2013) Dalton Trans 42:3029
10. Hille R, Hall J, Basu P (2014) Chem Rev 114:3963
11. Maia L, Moura I, Moura JJG (2017) Molybdenum and tungsten-containing enzymes: an overview. In: Hille R, Schulzke C, Kirk M (eds) Molybdenum and tungsten enzymes: biochemistry. RSC Metallobiology Series No. 5. The Royal Society of Chemistry, Cambridge, pp 1–80. doi:10.1039/9781782623915-00001
12. Anbar AD, Knoll AH (2002) Science 297:1137
13. Zerkle AL, House CH, Cox RP, Canfield DE (2006) Geobiology 4:285
14. Scott C, Lyons TW, Bekker A, Shen Y, Poulton SW, Chu X, Anbar AD (2008) Nature 452:456
15. Glass JB, Wolfe-Simon F, Anbar AD (2009) Geobiology 7:100
16. Schoepp-Cothenet B, van Lis R, Philippot P, Magalon A, Russell MJ, Nitschke W (2012) Sci Rep 2:263
17. Zhang X, Sigman DM, Morel FMM, Kraepiel AML (2014) Proc Natl Acad Sci U S A 111:4782

18. Maia LB, Moura JJG (2011) J Biol Inorg Chem 16:443
19. Maia LB, Moura JJG (2014) Chem Rev 114:5273
20. Sparacino-Watkins CE, Tejero J, Sun B, Gauthier MC, Thomas J, Ragireddy V, Merchant BA, Wang J, Azarov I, Basu P, Gladwin MT (2014) J Biol Chem 289:10345
21. Maia LB, Moura JJG (2015) J Biol Inorg Chem 20:403
22. Maia LB, Pereira V, Mira L, Moura JJG (2015) Biochemistry 54:685
23. Wang J, Krizowski S, Fischer K, Niks D, Tejero J, Sparacino-Watkins C, Wang L, Ragireddy P, Frizzell S, Kelley EE, Zhang Y, Basu P, Hille R, Schwarz G, Gladwin MT (2015) Antioxid Redox Signal 23:283
24. Grimaldi S, Schoepp-Cothenet B, Ceccaldi P, Guigliarelli B, Magalon A (2013) Biochim Biophys Acta 1827:1048
25. Hartmann T, Schwanhold N, Leimkühler S (2014) Biochim Biophys Acta 1854:1090
26. Maia LB, Moura JJG, Moura I (2015) J Biol Inorg Chem 20:287
27. Akaba S, Seo M, Dohmae N, Takio K, Sekimoto H, Kamiya Y, Furuya N, Komano T, Koshiba T (1999) J Biochem 126:395
28. Seo M, Koiwai H, Akaba S, Komano T, Oritani T, Kamiya Y, Koshiba T (2000) Plant J 23:481
29. Seo M, Peeters AJ, Koiwai H, Oritani T, Marion-Poll A, Zeevaart JA, Koornneef M, Kamiya Y, Koshiba T (2000) Proc Natl Acad Sci U S A 97:12908
30. Huang DY, Furukawa A, Ichikawa Y (1999) Arch Biochem Biophys 364:264
31. Hille R (2005) Arch Biochem Biophys 433:107
32. Garattini E, Fratelli M, Terao M (2008) Cell Mol Life Sci 65:1019
33. Pryde DC, Dalvie D, Hu Q, Jones P, Obach RS, Tran TD (2010) J Med Chem 53:8441
34. Garattini E, Terao M (2011) Drug Metab Rev 43:374
35. Hille R, Nishino T, Bittner F (2011) Coord Chem Rev 255:1179
36. Swenson TL, Casida JE (2013) Toxicol Sci 133:22
37. Terao M, Romão MJ, Leimkühler S, Bolis M, Fratelli M, Coelho C, Santos-Silva T, Garattini E (2016) Arch Toxicol 90:753
38. Hille R, Nishino T (1995) FASEB J 9:995
39. Hesberg C, Haensch R, Mendel RR, Bittner F (2004) J Biol Chem 279:13547
40. Yesbergenova Z, Yang G, Oron E, Soffer D, Flur R, Sagi M (2005) Plant J 42:862
41. Hille R (2006) Eur J Inorg Chem 2006:1913
42. Nishino T, Okamoto K, Eger BT, Pai EF, Nishino T (2008) FEBS J 275:3278
43. Zarepour M, Kaspari K, Stagge S, Rethmeier R, Mendel RR, Bittner F (2010) Plant Mol Biol 72:301
44. Okamoto K, Kusano T, Nishino T (2013) Curr Pharm Des 19:2606
45. Kappler U, Bennett B, Rethmeier J, Schwarz G, Deutzmann R, McEwan AG, Dahl C (2000) J Biol Chem 275:13202
46. Di Salle A, D'Errico G, La Cara F, Cannio R, Rossi M (2006) Extremophiles 10:587
47. Denger K, Weinitschke S, Smits THM, Schleheck D, Cook AM (2008) Microbiology 154:256
48. Wilson JJ, Kappler U (2009) Biochim Biophys Acta 1787:1516
49. Vijayakumar K, Gunny R, Grunewald S, Carr L, Chong KW, DeVile C, Robinson R, McSweeney N, Prabhakar P (2001) Pediatr Neurol 45:246
50. Johnson JL (2003) Prenat Diagn 23:6
51. Sass JO, Gunduz A, Araujo Rodrigues Funayama C, Korkmaz B, Dantas Pinto KG, Tuysuz B, Yanasse Dos Santos L, Taskiran E, de Fátima Turcato M, Lam CW, Reiss J, Walter M, Yalcinkaya C, Camelo Junior JS (2010) Brain Dev 32:544
52. Carmi-Nawi N, Malinger G, Mandel H, Ichida K, Lerman-Sagie T, Lev D (2011) J Child Neurol 26:460
53. Schwarz G, Belaidi A (2013) Met Ions Life Sci 13:415
54. Schwarz G (2016) Curr Opin Chem Biol 31:179
55. Fraústo da Silva JJR, Williams RJP (2001) The biological chemistry of the elements – the inorganic chemistry of life. Oxford University Press, Oxford
56. Bjornsson R, Lima FA, Spatzal T, Weyhermueller T, Glatzel P, Einsle O, Neese F, DeBeer S (2014) Chem Sci 5:3096

57. Bjornsson R, Neese F, Schrock RR, Einsle O, DeBeer S (2015) J Biol Inorg Chem 20:447
58. Burgmayer SJN, Stiefel EI (1985) J Chem Educ 62:943
59. Harlan EE, Berg JM, Holm RH (1986) J Am Chem Soc 108:6992
60. Holm RH, Berg JM (1986) Acc Chem Res 19:363
61. Holm RH (1987) Chem Rev 87:1401
62. Holm RH (1990) Coord Chem Rev 100:183
63. Xiao Z, Young CG, Enemark JH, Wedd AG (1992) J Am Chem Soc 114:9194
64. Holm RH, Donahue JP (1993) Polyhedron 12:571
65. Scbultz BE, Gbeller SF, Muetterties MC, Scott MJ, Holm RH (1993) J Am Chem Soc 115:2714
66. Holm RH, Kennepohl P, Solomon EI (1996) Chem Rev 96:2239
67. Donahue JP (2006) Chem Rev 106:4747
68. Reichenbecher W, Schink B (1999) Biochim Biophys Acta 1430:245
69. Prisner T, Lyubenova S, Atabay Y, MacMillan F, Kroger A, Klimmek O (2003) J Biol Inorg Chem 8:419
70. Nagarajan K, Joshi HK, Chaudhury PK, Pal K, Cooney JJA, Enemark JH, Sarkar S (2004) Inorg Chem 43:4532
71. Yang J, Rothery R, Sempombe J, Weiner JH, Kirk ML (2009) J Am Chem Soc 131:15612
72. Pushie MJ, Doonan CJ, Moquin K, Weiner JH, Rothery R, George GN (2011) Inorg Chem 50:732
73. Giles LJ, Ruppelt C, Yang J, Mendel RR, Bittner F, Kirk ML (2014) Inorg Chem 53:9460
74. Hanzelmann P, Dobbek H, Gremer L, Huber R, Meyer O (2000) J Mol Biol 301:1221
75. Dobbek H, Gremer L, Kiefersauer R, Huber R, Meyer O (2002) Proc Natl Acad Sci U S A 99:15971
76. Resch M, Dobbek H, Meyer O (2005) J Biol Inorg Chem 10:518
77. Stein BW, Kirk ML (2014) Chem Commun 50:1104
78. Hille R, Dingwall S, Wilcoxen J (2015) J Biol Inorg Chem 20:243
79. George GN, Pickering IJ, Yu EY, Prince RC, Bursakov SA, Gavel OY, Moura I, JJG M (2000) J Am Chem Soc 122:8321
80. Bursakov SA, Gavel OY, Di Rocco G, Lampreia J, Calvete J, Pereira AS, Moura JJ, Moura I (2004) J Inorg Biochem 98:833
81. Pauleta SR, Duarte AG, Carepo MS, Pereira AS, Tavares P, Moura I, Moura JJG (2007) Biomol NMR Assign 1:81
82. Rivas MG, Carepo MS, Mota CS, Korbas M, Durand MC, Lopes AT, Brondino CD, Pereira AS, George GN, Dolla A, Moura JJ, Moura I (2009) Biochemistry 48:873
83. Cvetkovic A, Menon AL, Thorgersen MP, Scott JW, Poole FL II, Jenney FE, Lancaster WA, Praissman JL, Shanmukh S, Vaccaro BJ, Trauger SA, Kalisiak E, Apon JV, Siuzdak G, Yannone SM, Tainer JA, Adams MWW (2010) Nature 466:779
84. Carepo MS, Pauleta SR, Wedd AG, Moura JJG, Moura I (2014) J Biol Inorg Chem 19:605
85. Enroth C, Eger BT, Okamoto K, Nishino T, Nishino T, Pai EF (2000) Proc Natl Acad Sci U S A 97:10723
86. Okamoto K, Matsumoto K, Hille R, Eger BT, Pai EF, Nishino T (2004) Proc Natl Acad Sci U S A 101:7931
87. Pauff JM, Zhang J, Bell CE, Hille R (2008) J Biol Chem 283:4818
88. Cao H, Hall J, Hille R (2014) Biochemistry 53:533
89. Stein BW, Kirk ML (2015) J Biol Inorg Chem 20:183
90. Massey V, Edmondson D (1970) J Biol Chem 245:6595
91. Malthouse JPG, Bray RC (1980) Biochem J 191:265
92. Mendel RR, Kruse T (2012) Biochem Biophys Acta 1823:1568
93. Mendel R (2013) J Biol Chem 288:13165
94. Mendel RR, Leimkuhler S (2015) J Biol Inorg Chem 20:337
95. Murray KN, Watson JG, Chaykin S (1966) J Biol Chem 241:4798
96. Hille R, Sprecher H (1987) J Biol Chem 262:10914

97. Xia M, Dempski R, Hille R (1999) J Biol Chem 274:3323
98. Stockert AL, Shinde S, Anderson RF, Hille R (2002) J Am Chem Soc 124:14554
99. Laughlin LJ, Young CG (1996) Inorg Chem 35:1050
100. Enemark JH, Cooney JJA, Wang J-J, Holm RH (2004) Chem Rev 104:1175
101. Sugimoto H, Tsukube H (2008) Chem Soc Rev 37:2609
102. Basu P, Burgmayer SJN (2015) J Biol Inorg Chem 20:373
103. Bray RC, Meriwether LS (1966) Nature 212:467
104. George GN, Bray RC (1988) Biochemistry 27:3603
105. Palmer G, Bray RC, Beinerrt H (1964) J Biol Chem 239:2657
106. Bray RC, Palmer G, Beinert H (1964) J Biol Chem 239:2667
107. Bray RC, Vanngard T (1969) Biochem J 114:725
108. Tanner S, Bray RC, Bergmann F (1978) Biochem Soc Trans 6:1328
109. Bray RC, George GN (1985) Biochem Soc Trans 13:560
110. Bray RC, Knowles PF, Meriwether LS (1967) ESR and the role of molybdenum in enzymic catalysis by xanthine oxidase. In: Ehrenberg A, Malmstrom BG, Vänngård T (eds) Magnetic resonance in biological systems. Pergamon Press, London
111. Edmondson DE, Ballou D, van Heuvelen A, Palmer G, Massey V (1973) J Biol Chem 248:6135
112. Hawkes TR, George GN, Bray RC (1984) Biochem J 218:961
113. McWhirter RB, Hille R (1991) J Biol Chem 266:23724
114. Gutteridge S, Bray RC (1980) Biochem J 189:615
115. Malthouse JPG, George GN, Lowe DJ, Bray RC (1981) Biochem J 197:421
116. Hanson GR, Wilson GL, Bailey TD, Pilbrow JR, Wedd AG (1987) J Am Chem Soc 109:2609
117. Massey V, Komai H, Palmer G, Elion GB (1970) J Biol Chem 246:2837
118. Drew SC, Hanson GR (2009) Inorg Chem 48:2224
119. Drew SC, Hill JP, Lane I, Hanson GR, Gable RW, Young CG (2007) Inorg Chem 46:2373
120. Drew SC, Young CG, Hanson GR (2007) Inorg Chem 46:2388
121. Bordas J, Bray RC, Garner CD, Gutteridge S, Hasnain S (1980) Biochem J 191:499
122. Malthouse JPG, George GN, Lowe DJ, Bray RC (1981) Biochem J 199:629
123. Wilson GL, Greenwood RJ, Pilbrow JR, Spence JT, Wedd AG (1991) J Am Chem Soc 113:6803
124. Cramer SP, Johnson JL, Rajagopalan KV, Sorrell TN (1979) Biochem Biophys Res Commun 91:434
125. Goodman BA, Raynor JB (1970) Adv Inorg Chem Radiochem 13:135
126. Hille R, Massey V (1985) Molybdenum-containing hydroxylases: xanthine oxidase aldehyde oxidase and sulfite oxidase. In: Spiro TG (ed) Molybdenum enzymes. Wiley, New York, NY, pp 443–518
127. Tullius TD, Kurtz DM Jr, Conradson SD, Hodgson KO (1979) J Am Chem Soc 101:2776
128. Howes BD, Bray RC, Richards RL, Turner NA, Bennett B, Lowe DJ (1996) Biochemistry 35:1432
129. Manikandan P, Choi E-Y, Hille R, Hoffman BM (2001) J Am Chem Soc 123:2658
130. Truglio JJ, Theis K, Leimkuhler S, Rappa R, Rajagopalan KV, Kisker C (2002) Structure 10:115
131. Gutteridge S, Tanner SJ, Bray RC (1978) Biochem J 175:869
132. Bray RC, Barber MJ, Lowe DJ (1978) Biochem J 171:653
133. Coughlan MP, Rajagopalan KV, Handler P (1969) J Biol Chem 244:2658
134. Pick FM, Bray RC (1969) Biochem J 114:735
135. Kanda M, Rajagopalan KV (1972) J Biol Chem 247:2177
136. Swann JC, Bray RC (1972) Eur J Biochem 26:407
137. Johnson JL, Waud WR, Cohen HJ, Rajagopalan KV (1974) J Biol Chem 249:5056
138. Gutteridge S, Tanner SJ, Bray RC (1978) Biochem J 175:887
139. Barber MJ, Coughlan MP, Kanda M, Rajagopalan KV (1980) Arch Biochem Biophys 201:468

140. Coughlan MP, Johnson JL, Rajagopalan KV (1980) J Biol Chem 255:2694
141. Maia L, Duarte RO, Ponces-Freire A, Moura JJG, Mira L (2007) J Biol Inorg Chem 12:777
142. Hille R, Kim JH, Hemann C (1993) Biochemistry 32:3973
143. Kim JH, Hille R (1993) J Biol Chem 268:44
144. Gutteridge S, Malthouse JPG, Bray RC (1979) J Inorg Biochem 11:355
145. Bray RC, Gutteridge S (1982) Biochemistry 21:5992
146. Morpeth FF, George GN, Bray RC (1984) Biochem J 220:235
147. Greenwood RJ, Wilson GL, Pilbrow JR, Wedd AG (1993) J Am Chem Soc 115:5385
148. Wilson GL, Kony M, Tiekink ERT, Pilbrow JR, Spence JT, Wedd AG (1988) J Am Chem Soc 110:6923
149. Bray RC, Knowles PF (1968) Proc R Soc Lond A 302:351
150. Malthouse JP, Bray RC (1983) Biochem J 215:101
151. Dowerah D, Spence JT, Singh R, Wedd AW, Wilson GL, Farchione F, Enemark JH, Kristofzski J, Bruckl M (1987) J Am Chem Soc 109:5655
152. Cammack R, Barber MJ, Bray RC (1976) Biochem J 157:469
153. Olson JS, Ballou DP, Palmer G, Massey V (1974) J Biol Chem 249:4363
154. Palmer G, Massey V (1969) J Biol Chem 244:2614
155. Barber MJ, Bray RC, Cammack R, Coughlan MP (1977) Biochem J 163:279
156. Bray RC, Knowles PF, Pick FM, Vänngård T (1968) Biochem J 107:601
157. Edmondson D, Massey V, Palmer G, Beacham LM, Elion GB (1972) J Biol Chem 247:1597
158. Pick FM, McGartoll MA, Bray RC (1971) Eur J Biochem 18:65
159. Howes BD, Pinhal NM, Turner NA, Bray RC, Anger G, Ehrenberg A, Raynor JB, Lowe DJ (1990) Biochemistry 29:6120
160. Shanmugam M, Zhang B, McNaughton RL, Kinney RA, Hille R, Hoffman BM (2010) J Am Chem Soc 132:14015
161. Lowe DJ, Barber MJ, Pawlik RT, Bray RC (1976) Biochem J 155:81
162. Hille R, Stewart RC, Fee JA, Massey V (1983) J Biol Chem 258:4849
163. Stewart RC, Hille R, Massey V (1984) J Biol Chem 259:14426
164. Barber MJ, Siegel LM (1983) Biochemistry 22:618
165. George GN, Bray RC (1983) Biochemistry 22:1013
166. Cramer SP, Hille R (1985) J Am Chem Soc 107:8164
167. Cao H, Hall J, Hille R (2011) J Am Chem Soc 133:12414
168. George GN, Bray RC (1983) Biochemistry 22:5443
169. Hille R, Hagen WR, Dunham WR (1985) J Biol Chem 260:10569
170. Lowe DJ, Lynden-Bell RM, Bray RC (1972) Biochem J 130:239
171. Lowe DJ, Bray RC (1978) Biochem J 169:471
172. Canne C, Lowe DJ, Fetzner S, Adams B, Smith AT, Kappl R, Bray RC, Hüttermann J (1999) Biochemistry 38:14077
173. Caldeira J, Belle V, Asso M, Guigliarelli B, Moura I, Moura JJG, Bertrand P (2000) Biochemistry 39:2700
174. More C, Asso M, Roger G, Guigliarelli B, Caldeira J, Moura J, Bertrand P (2005) Biochemistry 44:11628
175. Rupp H, Rao KK, Hall DO, Cammack R (1978) Biochim Biophys Acta 537:255
176. Barber MJ, Salerno JC, Siegel LM (1982) Biochemistry 21:1648
177. Coffman RE, Buettner GR (1979) J Phys Chem 83:2392
178. Moura JJG, Xavier AV, Bruschi M, Le Gall J, Hall DO, Cammack R (1976) Biochem Biophys Res Commun 72:782
179. Turner N, Barata B, Bray RC, Deistung J, Le Gall J, Moura JJ (1987) Biochem J 243:755
180. Barata BA, LeGall J, Moura JJ (1993) Biochemistry 32:11559
181. Romão MJ, Archer M, Moura I, Moura JJG, LeGall J, Engh R, Schneider M, Hof P, Huber R (1995) Science 270:1170
182. Huber R, Hof P, Duarte RO, Moura JJG, Moura I, Liu M-Y, Legall J, Hille R, Archer M, Romão MJ (1996) Proc Natl Acad Sci USA 93:8846

183. Rebelo JM, Dias JM, Huber R, Moura JJG, Romão MJ (2001) J Biol Inorg Chem 6:791
184. Krippahl L, Palma N, Moura I, Moura JJG (2006) Eur J Inorg Chem 19:3835
185. Barata BA, Liang J, Moura I, LeGall J, Moura JJG, Huynh BH (1992) Eur J Biochem 204:773
186. Duarte RO, Archer M, Dias JM, Bursakov S, Huber R, Moura I, Romão MJ, Moura JJG (2000) Biochem Biophys Res Commun 268:745
187. Santos-Silva T, Ferroni F, Thapper A, Marangon J, Gonzalez PJ, Rizzi AC, Moura I, Moura JJG, Romão MJ, Brondino CD (2009) J Am Chem Soc 131:7990
188. Mehra RK, Coughlan MP (1984) Arch Biochem Biophys 229:585
189. Coughlan MP, Mehra RK, Barber MJ, Siegel LM (1984) Arch Biochem Biophys 229:596
190. Boer DR, Thapper A, Brondino CD, Romão MJ, Moura JJG (2004) J Am Chem Soc 126:8614
191. Thapper A, Boer DR, Brondino CD, Moura JJG, Romao MJ (2007) J Biol Inorg Chem 12:353–366
192. Gladyshev VN, Khangulov SV, Stadtman TC (1994) Proc Natl Acad Sci USA 91:232
193. Schrader T, Rienhofer A, Andreesen JR (1999) Eur J Biochem 264:862
194. Self WT, Stadtman TC (2000) Proc Natl Acad Sci USA 97:7208
195. Wagener N, Pierik AJ, Ibdah A, Hille R, Dobbek H (2009) Proc Natl Acad Sci USA 106:11055
196. Hofmann M, Kassube JK, Graf T (2005) J Biol Inorg Chem 10:490
197. Siegbahn PEM, Shestakov AF (2005) J Comput Chem 26:888
198. Zhang B, Hemann CF, Hille R (2010) J Biol Chem 285:12571
199. Gourlay C, Nielsen DJ, White JM, Knottenbelt SZ, Kirk ML, Young CG (2006) J Am Chem Soc 128:2164
200. Wilcoxen J, Snider S, Hille R (2011) J Am Chem Soc 133:12934
201. Gnida M, Ferner R, Gremer L, Meyer O, Meyer-Klaucke W (2003) Biochemistry 42:222
202. Shanmugam M, Wilcoxen J, Habel-Rodriguez D, Cutsail GEI, Kirk ML, Hoffman BM, Hille R (2013) J Am Chem Soc 135:17775
203. Feng C, Tollin G, Enemark JH (2007) Biochim Biophys Acta 1774:527
204. Kappler U, Enemark JH (2015) J Biol Inorg Chem 20:253
205. Loschi L, Brokx SJ, Hills TL, Zhang G, Bertero MG, Lovering AL, Weiner JH, Strynadka NCJ (2004) J Biol Chem 279:50391
206. George GN, Doonan CJ, Rothery RA, Boroumand N, Weiner JH (2007) Inorg Chem 46:2
207. Havelius KGV, Reschke S, Horn S, Doerlng A, Niks D, Hille R, Schulzke C, Leimkuehler S, Haumann M (2011) Inorg Chem 50:741
208. Ott G, Havemeyer A, Clement B (2015) J Biol Inorg Chem 20:265
209. Kisker C, Schindelin H, Pacheco A, Wehbi WA, Garrett RM, Rajagopalan KV, Enemark JH, Rees DC (1997) Cell 91:973
210. Griffith OW (1987) Methods Enzymol 143:366
211. Pacheco A, Hazzard JT, Tollin G, Enemark JH (1999) J Biol Inorg Chem 4:390
212. Feng CJ, Kedia RV, Hazzard JT, Hurley JK, Tollin G, Enemark JH (2002) Biochemistry 41:5816
213. Johnson-Winters K, Nordstrom AR, Emesh S, Astashkin AV, Rajapakshe A, Berry RE, Tollin G, Enemark JH (2010) Biochemistry 49:1290
214. Hille R (1994) Biochim Biophys Acta 1184:143
215. Brody MS, Hille R (1995) Biochim Biophys Acta 1253:133
216. Pietsch MA, Hall MB (1996) Inorg Chem 35:1273
217. Brody MS, Hille R (1999) Biochemistry 38:6668
218. Thomson LM, Hall MB (2001) J Am Chem Soc 123:3995
219. Peariso K, McNaughton RL, Kirk ML (2002) J Am Chem Soc 124:9006
220. Wilson HL, Rajagopalan KV (2004) J Biol Chem 279:15105
221. Kail BW, Perez LM, Zaric SD, Millar AJ, Young CG, Hall MB, Basu P (2006) Chem Eur J 12:7501
222. Bailey S, Rapson T, Winters-Johnson K, Astashkin AV, Enemark JH, Kappler U (2009) J Biol Chem 284:2053
223. Byrne RS, Haensch R, Mendel RR, Hille R (2009) J Biol Chem 284:35479

224. Cohen HJ, Fridovich I, Rajagopalan KV (1971) J Biol Chem 246:374
225. Kessler DL, Rajagopalan KV (1972) J Biol Chem 247:6566
226. Kessler DL, Johnson JL, Cohen HJ, Rajagopalan KV (1974) Biochim Biophys Acta 334:86
227. Kessler DL, Rajagopalan KV (1974) Biochim Biophys Acta 370:389
228. Johnson JL, Rajagopalan KV (1976) J Biol Chem 251:5505
229. Gutteridge S, Lamy MT, Bray RC (1980) Biochem J 191:285
230. Lamy MT, Gutteridge S, Bary RC (1980) Biochem J 185:397
231. Bray RC, Lamy MT, Gutteridge S, Wilkinson T (1982) Biochem J 201:241
232. Astashkin AV, Raitsimring AM, Feng CJ, Johnson JL, Rajagopalan KV, Enemark JH (2002) J Am Chem Soc 124:6109
233. Garton SD, Garrett RM, Rajagopalan KV, Johnson MK (1997) J Am Chem Soc 119:2590
234. Astashkin AV, Johnson-Winters K, Klein EL, Feng C, Wilson HL, Rajagopalan KV, Raitsimring AM, Enemark JH (2008) J Am Chem Soc 130:8471
235. Enemark JH, Raitsimring AM, Astashkin AV, Klein EL (2011) Faraday Discuss 148:249
236. Klein EL, Raitsimring AM, Astashkin AV, Rajapakshe A, Johnson-Winters K, Arnold AR, Potapov A, Goldfarb D, Enemark JH (2012) Inorg Chem 51:1408
237. Bray RC, Gutteridge S, Lamy MT, Wilkinson T (1983) Biochem J 211:227
238. Astashkin AV, Klein EL, Enemark JH (2007) J Inorg Biochem 101:1623
239. Doonan CJ, Wilson HL, Bennett B, Prince RC, Rajagopalan KV, George GN (2008) Inorg Chem 47:2033
240. Klein EL, Astashkin AV, Ganyushin D, Riplinger C, Johnson-Winters K, Neese F, Enemark JH (2009) Inorg Chem 48:4743
241. Karakas E, Wilson HL, Graf TN, Xiang S, Jaramillo-Buswuets S, Rajagopalan KV, Kisker C (2005) J Biol Chem 280:33506
242. Pushie MJ, Doonan CJ, Wison HL, Rajagopalan KV, George GN (2011) Inorg Chem 50:9406
243. Rajapakshe A, Johnson-Winters K, Nordstrom AR, Meyers KT, Emesh S, Astashkin AV, Enemark JH (2010) Biochemistry 49:5154
244. Astashkin AV, Johnson-Winters K, Klein EL, Byrne RS, Hille R, Raitsimring AM, Enemark JH (2007) J Am Chem Soc 129:14800
245. Astashkin AV, Hood BL, Feng C, Hille R, Mendel RR, Raitsimring AM, Enemark JH (2005) Biochemistry 44:13274
246. Doonan CJ, Wilson HL, Garrett RM, Bennet B, Prince RC, Rajagopalan KV, George GM (2007) J Am Chem Soc 129:9421
247. Garrett RM, Rajagopalan KV (1996) J Biol Chem 271:7387
248. George GN, Garrett RM, Prince RC, Rajagopalan KV (1996) J Am Chem Soc 118:8588
249. Qiu JA, Wilson HL, Pushie MJ, Kisker C, George GN, Rajagopalan KV (2010) Biochemistry 49:3989
250. George GN, Garrett RM, Prince RC, Rajagopalan KV (2004) Inorg Chem 43:8456
251. Reschke S, Niks D, Wilson H, Sigfridsson KGV, Haumann M, Rajagopalan KV, Hille R, Leimkuhler S (2013) Biochemistry 52:8295
252. Raitsimring AM, Pacheco A, Enemark JH (1998) J Am Chem Soc 120:11263
253. Astashkin AV, Mader ML, Pacheco A, Enemark JH, Raitsimring AM (2000) J Am Chem Soc 122:5294
254. George GN (1985) J Magn Reson 64:384
255. Astashkin AV, Feng CJ, Raitsimring AM, Enemark JH (2005) J Am Chem Soc 127:502
256. Astashkin AV, Neese F, Raitsimring AM, Cooney JJA, Bultman E, Enemark JH (2005) J Am Chem Soc 127:16713
257. Peariso K, Chohan BS, Carrano CJ, Kirk ML (2003) Inorg Chem 42:6194
258. George GN, Prince RC, Kipke CA, Sunde RA, Enemark JH (1988) Biochem J 256:307
259. Pacheco A, Basu P, Borbat P, Raitsimring AM, Enemark JH (1996) Inorg Chem 35:7001
260. George GN, Garrett RM, Graf T, Prince RC, Rajagopalan KV (1998) J Am Chem Soc 120:4522
261. Schink B (1985) Arch Microbiol 142:295

262. Messerschmidt A, Niessen H, Abt D, Einsle O, Schink B, Kroneck PM (2004) Proc Natl Acad Sci U S A 101:11571
263. Moura JJ, Brondino CD, Trincao J, Romao MJ (2004) J Biol Inorg Chem 9:791
264. Jormakka M, Yokoyama K, Yano T, Tamakoshi M, Akimoto S, Shimamura T, Curmi P, Iwata S (2008) Nat Struct Mol Biol 15:730
265. Rothery RA, Workun GJ, Weiner JH (2008) Biochim Biophys Acta 1778:1897
266. Ellis PJ, Conrads T, Hille R, Kuhn P (2001) Structure 9:125
267. Conrads T, Hemann C, George GN, Pickering IJ, Prince RC, Hille R (2002) J Am Chem Soc 124:11276
268. Silver S, Phung LT (2005) Appl Environ Microbiol 71:599
269. Warelow TP, Oke M, Schoepp-Cothenet B, Dahl JU, Bruselat N, Sivalingam GN, Leimkuhler S, Thalassinos K, Kappler U, Naismith JH, Santini JM (2013) PLoS One 8:e72535
270. Potter L, Angove H, Richardson D, Cole J (2001) Adv Microb Physiol 45:51
271. Stolz JF, Basu P (2002) ChemBioChem 3:198
272. Bertero MG, Rothery RA, Palak M, Hou C, Lim D, Blasco F, Weiner JH, Strynadka NCJ (2003) Nat Struct Biol 10:681
273. Jormakka M, Richardson D, Byrne B, Iwata S (2004) Structure 12:95
274. González PG, Correia C, Moura I, Brondino CD, Moura JJG (2006) J Inorg Biochem 100:1015
275. Martinez-Espinosa RM, Dridge EJ, Bonete MJ, Butt JN, Butler CS, Sargent F, Richardson DJ (2007) FEMS Microbiol Lett 276:129
276. Kraft B, Strous M, Tegetmeyer HE (2011) J Biotechnol 155:104
277. Gonzalez PJ, Mota CS, Brondino CD, Moura I, Moura JJG (2013) Coord Chem Rev 257:315
278. Sparacino-Watkins C, Stolzb JF, Basu P (2014) Chem Soc Rev 43:676
279. Bertero MG, Rothery RA, Boroumand N, Palak M, Blasco F, Ginet N, Weiner JH, Strynadka NCJ (2005) J Biol Chem 280:14836
280. Dias JM, Than ME, Humm A, Huber R, Bourenkov GP, Bartunik HD, Bursakov S, Calvete J, Caldeira J, Carneiro C, Moura JJG, Moura I, Romão MJ (1999) Structure 7:65
281. Najmudin S, Gonzalez PJ, Trincao J, Coelho C, Mukhopadhyay A, Cerqueira N, Romão CC, Moura I, Moura JJG, Brondino CD, Romão MJ (2008) J Biol Inorg Chem 13:773
282. Coelho C, Gonzalez PJ, Moura JJG, Moura I, Trincao J, Romão MJ (2011) J Mol Biol 408:932
283. Coelho C, Gonzalez PJ, Trincão J, Carvalho AL, Najmudin S, Hettman T, Dieckman S, Moura JJG, Moura I, Romão MJ (2007) Acta Crystallogr F 63:516
284. Jepson BJN, Mohan S, Clarke TA, Gates AJ, Cole JA, Butler CS, Butt JN, Hemmings AM, Richardson DJ (2007) J Biol Chem 282:6425
285. Arnoux P, Sabaty M, Alric J, Frangioni B, Guigliarelli B, Adriano JM, Pignol D (2003) Nat Struct Biol 10:928
286. Jepson BJN, Anderson LJ, Rubio LM, Taylor CJ, Butler CS, Flores E, Herrero A, Butt JN, Richardson DJ (2004) J Biol Chem 279:32212
287. Trchounian K, Poladyan A, Vassilian A, Trchounian A (2012) Crit Rev Biochem Mol Biol 47:236
288. Bagramyan K, Trchounian A (2003) Biochem Mosc 68:1159
289. Jones RW, Lamont A, Garland PB (1980) Biochem J 190:79
290. Berg BL, Li J, Heider J, Stewart V (1991) J Biol Chem 266:22380
291. Blasco F, Guigliarelli B, Magalon A, Asso M, Giordano G, Rothery RA (2001) Cell Mol Life Sci 58:179
292. Jormakka M, Tornroth S, Byrne B, Iwata S (2002) Science 295:1863
293. Richardson D, Sawers G (2002) Science 295:1842
294. Jormakka M, Byrne B, Iwata S (2003) Curr Opin Struct Biol 13:418
295. Jormakka M, Byrne B, Iwata S (2003) FEBS Lett 545:25
296. Sawers G, Heider J, Zehelein E, Bock A (1991) J Bacteriol 173:4983
297. Pommier J, Mandrand MA, Holt SE, Boxer DH, Giodano G (1992) Biochim Biophys Acta 1107:305

298. Sawers G (1994) Antonie Van Leeuwenhoek 66:57
299. Abaibou H, Pommier J, Benoit S, Giordano G, Mandrandberthelot MA (1995) J Bacteriol 177:7141
300. Benoit S, Abaibou H, Mandrand-Berthelot MA (1998) J Bacteriol 180:6625
301. Friedebold J, Bowien B (1993) J Bacteriol 175:4719
302. Friedebold J, Mayer F, Bill E, Trautwein AX, Bowien B (1995) Biol Chem Hoppe Seyler 376:561
303. Boyington JC, Gladyshev VN, Khangulov SV, Stadtman TC, Sun PD (1997) Science 275:1305
304. Costa C, Teixeira M, LeGall J, Moura JJG, Moura I (1997) J Biol Inorg Chem 2:198
305. Oh JI, Bowien B (1998) J Biol Chem 273:26349
306. Raaijmakers HCA, Romão MJ (2006) J Biol Inorg Chem 11:849
307. Rivas M, Gonzalez P, Brondino CD, Moura JJG, Moura I (2007) J Inorg Biochem 101:1617
308. Thome R, Gust A, Toci R, Mendel R, Bittner F, Magalon A, Walburger A (2012) J Biol Chem 287:4671
309. Hartmann T, Leimkuhler S (2013) FEBS J 280:6083
310. Gladyshev VN, Khangulov SV, Axley MJ, Stadtman TC (1994) Proc Natl Acad Sci U S A 91:7708
311. Gladyshev VN, Boyington JC, Khangulov SV, Grahame DA, Stadtman TC, Sun PD (1996) J Biol Chem 271:8095
312. Khangulov SV, Gladyshev VN, Dismukes GC, Stadtman TC (1998) Biochemistry 37:3518
313. George GN, Colangelo CM, Dong J, Scott RA, Khangulov SV, Gladyshev VN, Stadtman TC (1998) J Am Chem Soc 120:1267
314. Niks D, Duvvuru J, Escalona M, Hille R (2016) J Biol Chem 291:1162
315. Maia LB, Fonseca L, Moura I, Moura JJG (2016) J Am Chem Soc 138:8834
316. Raaijmakers H, Macieira S, Dias JM, Teixeira S, Bursakov S, Huber R, Moura JJG, Moura I, Romão MJ (2002) Structure 10:1261
317. George GN, Costa C, Moura JJG, Moura I (1999) J Am Chem Soc 121:2625
318. Jollie DR, Lipscomb JD (1991) J Biol Chem 266:21853
319. Brondino CD, Passeggi MCG, Caldeira J, Almendra MJ, Feio MJ, Moura JJG, Moura I (2004) J Biol Inorg Chem 9:145
320. Hagedoorn PL, Hagen WR, Stewart LJ, Docrat A, Bailey S, Garner CD (2003) FEBS Lett 555:606
321. Gonzalez PJ, Rivas MG, Brondino CD, Bursakov SA, Moura I, Moura JJG (2006) J Biol Inorg Chem 11:609
322. Najmudin S, Gonzalez PJ, Trincao J, Coelho C, Mukhopadhyay A, Cerqueira N, Romão CC, Moura I, Moura JJG, Brondino CD, Romão MJ (2008) J Biol Inorg Chem 13:737
323. Bennett B, Berks BC, Ferguson SJ, Thomson AJ, Richardson DJ (1994) Eur J Biochem 226:789
324. Butler CS, Charnock JM, Bennett B, Sears HJ, Reilly AJ, Ferguson SJ, Garner CD, Lowe DJ, Thomson AJ, Berks BC, Richardson DJ (1999) Biochemistry 38:9000
325. Butler CS, Charnock JM, Garner CD, Thomson AJ, Ferguson SJ, Berks BC, Richardson DJ (2000) Biochem J 352:859
326. Butler CS, Fairhurst SA, Ferguson SJ, Thomson AJ, Berks BC, Richardson DJ, Lowe DJ (2002) Biochem J 363:817
327. Bertrand P, Frangioni B, Dementin S, Sabaty M, Arnoux P, Guigliarelli B, Pignol D, Leger C (2007) J Phys Chem B 111:10300
328. Fourmond V, Burlat B, Dementin S, Arnoux P, Sabaty M, Boiry S, Guigliarelli B, Bertrand P, Pignol D, Leger C (2008) J Phys Chem B 112:15478
329. Simpson PJL, McKinzie AA, Codd R (2010) Biochem Biophys Res Commun 398:13
330. Simpson PJL, Richardson DJ, Codd R (2010) Microbiology 156:302
331. Biaso F, Burlat B, Guigliarelli B (2012) Inorg Chem 51:3409
332. Nemykin VN, Basu P (2003) Inorg Chem 42:4046

333. Gangeswaran R, Lowe DJ, Eady RR (1993) Biochem J 289:335
334. Bray RC, Vincent SP, Lowe DJ, Clegg RA, Garland PB (1976) Biochem J 155:201
335. Clegg RA (1976) Biochem J 153:533
336. Vincent SP, Bray RC (1978) Biochem J 171:639
337. Godfrey C, Greenwood C, Thomson AJ, Bray RC, George GN (1984) Biochem J 224:601
338. George GN, Bray RC, Morpeth FF, Boxer DH (1985) Biochem J 227:925
339. Turner N, Ballard AL, Bray RC, Ferguson SJ (1988) Biochem J 252:925
340. Magalon A, Asso M, Guigliarelli B, Rothery RA, Bertrand P, Giordano G, Blasco F (1998) Biochemistry 37:7363
341. Hettmann T, Anemuller S, Borcherding H, Mathe L, Steinrucke P, Diekmann S (2003) FEBS Lett 534:143
342. Correia C, Besson S, Brondino CD, Gonzalez PJ, Fauque G, Lampreia J, Moura I, Moura JJG (2008) J Biol Inorg Chem 13:1321
343. Field SJ, Thornton NP, Anderson LJ, Gates AJ, Reilly A, Jepson BJN, Richardson DJ, George SJ, Cheesman MR, Butt JN (2005) Dalton Trans:3580
344. Anderson LJ, Richardson DJ, Butt JN (2001) Biochemistry 40:11294
345. Anderson LJ, Richardson DJ, Butt JN (2000) Faraday Discuss 116:155
346. George GN, Turner NA, Bray RC, Morpeth FF, Boxer DH, Cramer SP (1989) Biochem J 259:693

CW and Pulse EPR of Cytochrome P450 to Determine Structure and Function

Joshua S. Harbort, James J. De Voss, Jeanette E. Stok, Stephen G. Bell, and Jeffrey R. Harmer

Abstract Cytochromes P450 (P450s) are a diverse class of biological monooxygenases found in a wide variety of organisms, known for their chemical versatility and reaction specificity. While an array of chemical techniques are available to study P450s, continuous wave (CW) and pulse electron paramagnetic resonance (EPR) spectroscopies provide unique insight into the structure and function of the protein by probing various paramagnetic states. In this review we will demonstrate how EPR techniques are used to reveal information about the arrangement and conformation of P450 electron-transport protein complexes, characterise the active site oxidation state and the interactions with substrates and inhibitors. In addition, when combined with sample preparation using cryoreduction and freeze-quench techniques, EPR can be used to characterise short-lived intermediates formed during the catalytic cycle.

Keywords Cytochrome P450 • CW EPR • Pulse EPR • ENDOR • HYSCORE • Cryoreduction • Freeze-quench • Catalytic cycle • Compond I

J.S. Harbort • J.R. Harmer (✉)
Centre for Advanced Imaging, The University of Queensland, St. Lucia, QLD 4072, Australia
e-mail: jeffrey.harmer@cai.uq.edu.au

J.J. De Voss (✉) • J.E. Stok
School of Chemistry and Molecular Biosciences, The University of Queensland, St. Lucia, QLD 4072, Australia
e-mail: j.devoss@uq.edu.au

S.G. Bell (✉)
Department of Chemistry, The University of Adelaide, Adelaide, SA 5005, Australia
e-mail: stephen.bell@adelaide.edu.au

G. Hanson, L. Berliner (eds.), *Future Directions in Metalloprotein and Metalloenzyme Research*, Biological Magnetic Resonance 33,
DOI 10.1007/978-3-319-59100-1_5

Introduction

Cytochromes P450 are a large super-family of heme-centred monooxygenases that activate and split molecular oxygen, usually with the incorporation of one of the oxygen atoms into a bound substrate and one into water,

$$RH + O_2 + 2e^- + 2H^+ \rightarrow ROH + H_2O$$

P450s are found in a wide range of organisms across all domains of life and are named after the distinctive Soret peak they exhibit at 450 nm in optical absorption spectroscopy when the active site ferrous heme is coordinated to carbon monoxide. P450s are classified into numerical families (e.g. CYP199), lettered subfamilies (e.g. CYP199A), and isoforms within each subfamily (e.g. CYP199A2) [1]. They are important enzymes with roles in xenobiotic and steroid metabolism among others. The eukaryotic P450s are generally membrane-bound and more difficult to study than soluble bacterial P450s which consequently have been the focus of many studies [1].

Among monooxygenases, P450s are capable of accepting a chemically diverse range of substrates, and can perform a wide range of often highly regio- and stereoselective substrate modifications, making them key targets for synthetic biology [2, 3]. Through many years of research, the basic P450 catalytic reaction mechanism has been determined and is shown in Fig. 1 [1, 4] (though alternative reaction cycles have been proposed [5, 6]). The majority of the intermediates of the proposed catalytic cycle are paramagnetic which makes EPR a key technique in functional studies of the enzyme [1–5].

The general catalytic cycle for P450s (Fig. 1) and related heme proteins is as follows [1, 7–13]:

1. State **A**. The P450 resting state. In most P450s, this state consists of a low-spin, water-ligated hexacoordinate Fe^{III} heme ($S = 1/2$), also known as an aqua-ferric heme.
2. **A** → **B**. Entry of the substrate. Generally, the resting state water is displaced by entry of the substrate although there are a few exceptions [8]. Binding of the substrate in the active site (but not directly to the Fe^{III} ion) results in a pentacoordinate complex in a high-spin $S = 5/2$ state.
3. **B** → **C**. The heme is reduced from the Fe^{III}, $S = 5/2$ state to a ferrous heme (Fe^{II}) state which is EPR silent, $S = 2$. Under physiological conditions the reduction of P450 is facilitated by redox proteins but this pathway can be manipulated with chemical reductants.
4. **C** → **D**. Coordination of molecular oxygen to the heme centre, to form a ferrous dioxy complex (Fe^{II}-O_2), which is hexacoordinate and EPR silent ($S = 0$).
5. **D** → **E**. Formation of the ferric peroxo intermediate, (Fe^{III}-O_2^-; $S = 1/2$) via the second reduction event.
6. **E** → **F**. A protonation event, forming the ferric hydroperoxy intermediate (Fe^{III}-OOH; $S = 1/2$). The proton is delivered to the active site with the aid of a proton delivery network [9, 10, 12].

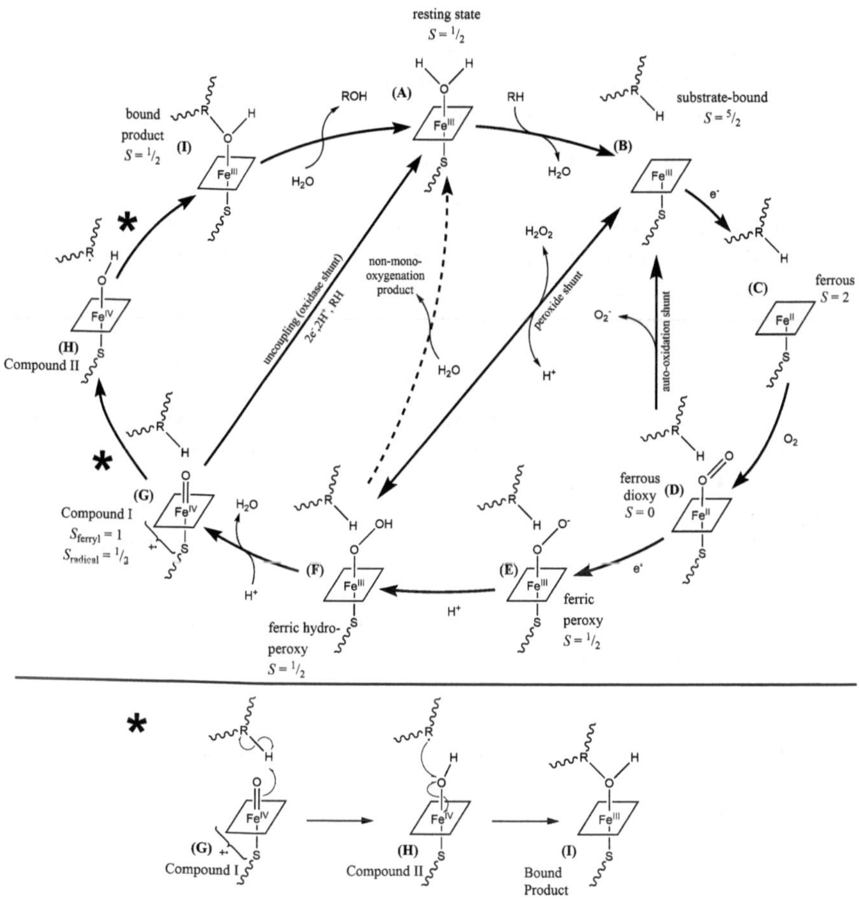

Fig. 1 Catalytic cycle for monooxygenation reactions catalysed by P450s [1, 2, 6]. The substrate molecule is represented by 'R'; the heme ring is represented by a square. The distal ligand of the heme is the thiol sulfur of a cysteine residue of the parent protein (represented by S and the wavy line). * indicates the radical rebound mechanism proposed by Groves et al. [95, 96] that occurs from intermediate **G** to **A**

7. **F → G**. Second protonation event, leading to O–O bond cleavage and the formation of the iron(IV) oxo (ferryl) porphyrin π-cation radical ($Fe^{IV} = O\ P^{+\bullet}$; $S^{total} = 1/2$) This is known as Compound I by analogy to the peroxidase catalytic cycle [11, 13].
8. **G → H**. Formation of compound II (Fe^{IV}-OH P, P indicates the porphyrin) by C–H bond scission, transfer of a hydrogen atom to the ferryl oxygen, and formation of a substrate radical (see rebound mechanism at the bottom of Fig. 1).
9. **H → I**. Oxygen rebound to form the C–OH bond in the product which can coordinate to the heme iron via the introduced oxygen (**I**).
10. **I → A**. At the end of the catalytic cycle, the bound product is released and water again coordinates to the resting state P450.

There are a number of other productive and unproductive (or 'uncoupling') reaction pathways which may occur during the cycle. Unproductive reaction pathways are generally measured by comparing the rate of co-oxidant consumption (e.g. NADH/NADPH) with the rate of product formation [1].

1. Auto-oxidation shunt, **D** → **B**. Consumption of one electron followed by release of $O_2^{-\bullet}$ [1, 14].
2. Peroxide shunt, **B** ↔ **F**. Utilises H_2O_2 (or other oxygen donors) to generate **F** from **B**, the ferric hydroperoxy intermediate (Fe^{III}-OOH) (or an analogue for other oxygen donors) [4, 15]. In the majority of P450 enzymes this is not a favoured mechanism because of oxidative heme degradation but the pathway has been used with *meta*-chloroperbenzoic acid (*m*-CPBA) to generate Compound I [13, 14]. A few P450s which can utilise this pathway have been identified [16–18]. The peroxide shunt pathway can also act in the direction **F** → **B**, consuming two electrons and releasing H_2O_2 and unmodified product.
3. Oxidase, **G** → **A**. Consumption of two electrons followed by release of H_2O and unmodified substrate from the active site instead of a modified substrate product [1, 19]. This occurs when the oxo ligand of compound I (**G**) is reduced instead of reacting with the substrate molecule.

Evidence from both P450s and related systems have suggested species **E**, **F** and **G** form a series of increasing chemical reactivity. Compound I, **G**, is the most active and has been shown to be responsible for C–H bond oxidations. The involvement of the two early intermediates in other P450 oxidations such as deformylation reactions (**E**), alkene epoxidation (**F**) and sulfoxidation (**F**) has been investigated (see section "Characterisation of the P450 Reaction Cycle by EPR").

P450s utilise a range of electron delivery systems, broadly divided into cytosolic (i.e. soluble) prokaryotic systems and membrane-bound eukaryotic systems [20]. Typically, electrons are sourced from NAD(P)H via a flavin adenine dinucleotide (FAD) containing reductase that then reduces a redox partner protein or cofactor consisting of either an iron-sulphur cluster or a flavin mononucleotide (FMN) redox centre which, in turn, transfers sequentially two electrons to the P450 active site. In type I P450s, the two electron transfer proteins are separate entities to the P450, whereas type II and type III have two, or all three partner domains, respectively, fused together into a single protein [20].

Since the resting state and many of the proposed intermediates of the P450 catalytic cycle are paramagnetic, electron paramagnetic resonance (EPR) is an important technique to determine structure and to monitor interactions of the active site with substrates and inhibitors. CW EPR spectroscopy at the conventional X-band frequencies (9–10 GHz) has sufficient resolution to enable the Fe^{III} heme *g*-factors of the paramagnetic centre to be measured accurately and is thus useful for investigating its electronic structure [7, 21–28]. Besides being diagnostic of the heme oxidation state, low-spin and high-spin CW EPR spectra are diagnostic of whether the water coordinated to the heme centre has been displaced by entry of a substrate

[21–24] and g-values have been shown in some cases to be sensitive to the binding of redox partners [29]. The information from CW EPR spectra is complementary to the information from the optical spectrum and both can be used to characterise different stages of the catalytic cycle [25], for example to identify the presence compound I [13]. Low-spin heme g-values are amenable to analysis using ligand field theory [26, 27] which allows an analysis of the distortion of the heme moiety [28]. CW EPR has also been used to investigate the iron heme active site of P450s within intact cells [7].

The CW EPR spectrum from a P450 enzyme in the Fe^{III} spin-state ($S = 1/2$ or $S = 5/2$) enables only the three principal g-values to be determined; no hyperfine couplings are resolved which would yield direct information on ligand coordination and the position of molecules in the active site. However, this information can be obtained with higher resolution EPR techniques such as electron spin echo envelope modulation (ESEEM) spectroscopy, pulsed and CW electron-nuclear double resonance (ENDOR) spectroscopy, and electron double resonance (ELDOR) detected NMR [30].

EPR based studies have been carried out to characterise the ferric resting state of the enzyme with water and substrate bound (Fig. 1, **A**, **B**). Characterisation of reaction intermediates by EPR is more demanding because of the fleeting nature of these species (Fig. 1, **D–I**) [9–12, 31–35]. Cryoreduction methodologies have made important contributions. In these experiments the dioxygen bound ferrous P450 heme centre is reduced (**D** → **E**) at low temperature using electrons derived from γ-ray exposure radiolysis [11], enabling an array of reactive species to be trapped at low temperature. These have been characterised through a combination of CW and pulse EPR studies, and their reactivity assessed by annealing the sample at elevated temperature to allow the reaction to proceed at a controlled rate. The chemical identities of an impressive number of these low temperature species have been definitively determined. This kinetic and structural information has provided convincing evidence for the correctness of the proposed catalytic cycle of the enzyme under physiological conditions [9–12, 31–35]. Key has been the application of rapid-freeze quench methods which finally (after 56 years of effort [36]) enabled P450 compound I to be characterised in detail [13, 14] and its reactivity compared to related enzymes and synthetic (model) systems [19, 33, 37, 38].

Interaction of the P450 with its redox partner can be studied using the technique of double electron-electron resonance (DEER, also known as pulsed electron double resonance, PELDOR) [39–46]. This technique allows the distance between two paramagnetic probes in the range *ca.* 15–80 Å to be determined [39].

This review starts with an elementary review of EPR spectroscopy methods, before discussing CW and pulsed EPR studies investigating the interaction of substrate and inhibitor molecules with the P450 resting state (species **A** and **B** in Fig. 1). Next, we describe methods used to isolate and characterise P450 reaction intermediates, and discuss their role in catalysis. We conclude by describing work on characterising P450 enzyme structural changes upon substrate binding and interaction with redox partners, using dipolar EPR (principally by DEER spectroscopy).

Basic EPR Theory for Cytochromes P450

EPR spectroscopy is concerned with the measurement of the energy levels of a paramagnetic centre, which delivers information on the electronic and geometric structure of the centre, and on the distance and orientation between centres for multi-spin systems. The general EPR Hamiltonian used to describe the spin system of a single paramagnetic centre in the laboratory frame can be written as [30].

$$H_0 = H_{EZ} + H_{ZFS} + H_{HF} + H_{NZ} + H_{NQ} \tag{1}$$

$$= \mu_B \vec{B}_0^T \mathbf{g} \vec{S} / \hbar + \vec{S}^T \mathbf{D} \vec{S} + \sum\nolimits_{i=1}^{m} \vec{S}^T \mathbf{A}_i \vec{I}_i - \mu_N \sum\nolimits_{i=1}^{m} g_{n,i} \vec{B}_0^T \vec{I}_i / \hbar + \sum\nolimits_{i=1; I>1/2}^{m} \vec{I}_i^T \mathbf{P}_i \vec{I}_i \tag{2}$$

Which contains the electron Zeeman (EZ), zero-field splitting (ZFS), hyperfine (HF) and nuclear quadrupole (NQ) interactions. $\vec{S} = \left[\hat{\mathbf{S}}_x, \hat{\mathbf{S}}_y, \hat{\mathbf{S}}_z\right]$ is a column vector of electron spin operators, and **I** a column vector of nuclear spin operators. If there is more than one paramagnetic spin, the magnetic interaction between them is described by

$$H_{ee} = \vec{S}_1^T \mathbf{D}_{ee} \vec{S}_2 \tag{3}$$

Tensor $\mathbf{D}_{ee}$ contains coupling parameters from the electron-electron dipole interaction used in dipolar EPR (e.g. using DEER) to determine the distance between the spins, and may contain an exchange coupling J if the wavefunctions of the spins overlap (as in Compound I, see Eq. (10)) [30].

The heme iron at the active site of a P450 in the Fe^{III} oxidation state can exist in either a low-spin (S = 1/2) or a high-spin (S = 5/2) state. In the low-spin state the EPR spectrum exhibits only the principal g-values—all magnetic interactions between the unpaired electron and nearby magnetic nuclei are smaller than the EPR spectrum linewidth and are thus not resolved. The CW EPR spectrum can thus be accurately simulated with a Hamiltonian consisting of only the electron Zeeman term ($\propto_B \vec{B}_0^T \mathbf{g} \vec{S} / \hbar$). Figure 2 (Table 1) shows X-band CW EPR simulations for a selection of typical low-spin and high-spin P450 spin-states using literature g-values in each case.

In the high-spin Fe^{III} case the zero-field splitting term ($\vec{S}^T \mathbf{D} \vec{S}$) also needs to be included—there are then $2S + 1$ electron spin manifolds and hence $2S$ possible EPR transitions between the manifolds. However, for a high-spin Fe^{III} heme centre the magnitude of the zero-field splitting is very large compared to the electron Zeeman term so that under the measurement conditions only the lowest ground-state Kramers doublet is populated and the system can thus be described in terms of an effective spin of S^{eff} = 1/2. For example, an S = 5/2 ion in tetragonal symmetry gives a CW EPR spectrum that can be described with spin S^{eff} = 1/2 where $g_\perp^{eff} \approx 6$ and $g_\parallel^{eff} \approx 2$. Figure 3 demonstrates how a simulation using effective g-values is related to one using the full spin Hamiltonian for an S = 5/2 spin state (computed with typical Fe^{III}

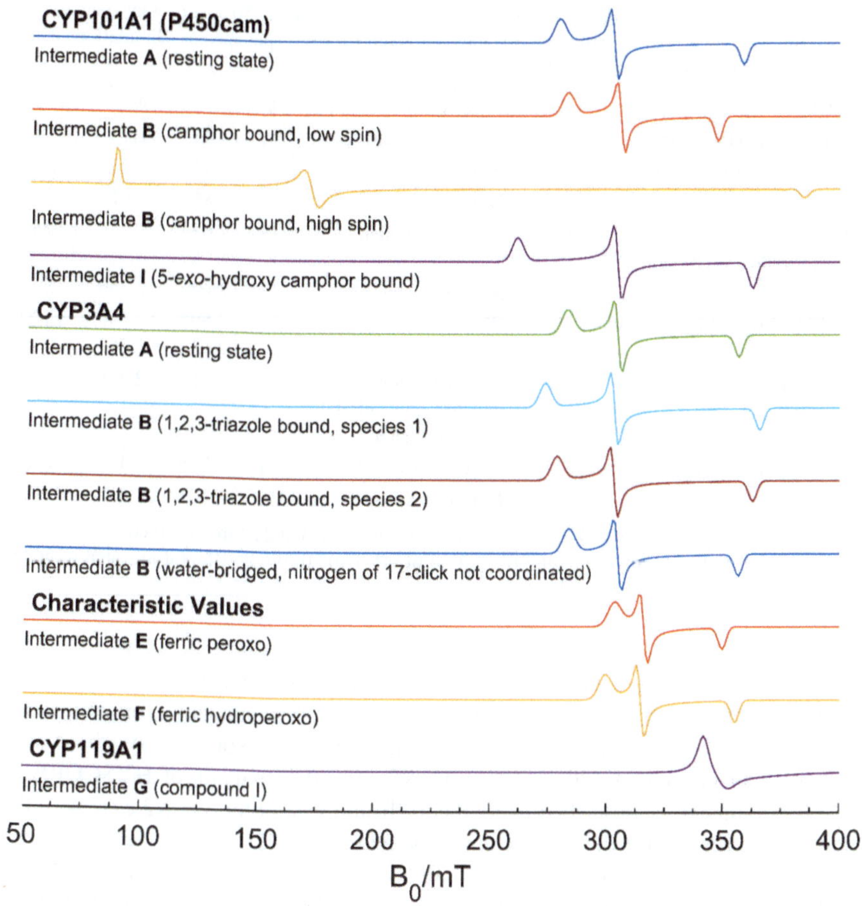

Fig. 2 Simulated X-band (9.6 GHz) CW EPR spectra for a selection of P450s and the intermediates **E**, **F**, **G** and **I** of the catalytic cycle. Table 1 lists the g-values used in the simulations. 17-click is defined in Fig. 8

heme parameters) [47–50]. A good description of high-spin EPR spectra with a biological inorganic emphasis is given by Telser et al. [51].

Within the lowest Kramers doublet the spin Hamiltonian of Eq. (2) can be recast as

$$H_0^{eff} = \mu_B \vec{B}_0^T \mathbf{g}^{eff} \vec{S} / \hbar + \sum\nolimits_{i=1}^{m} \vec{S}^T \mathbf{A}_i^{eff} \vec{I}_i - \mu_N \sum\nolimits_{i=1}^{m} \vec{B}_0^T \mathbf{g}_{n,i}{}^{eff} \vec{I}_i / \hbar + \sum\nolimits_{i=1, I>1/2}^{m} \vec{I}_i^T \mathbf{P}_i^{eff} \vec{I}_i, \quad (4)$$

where $\mathbf{g}^{\mathrm{eff}}$, $\mathbf{A}^{\mathrm{eff}}$, $\mathbf{g}_{\mathrm{n}}{}^{\mathrm{eff}}$ and $\mathbf{P}^{\mathrm{eff}}$ are effective coupling tensors in the lowest Kramers doublet. In the case of the $S = 5/2$ heme paramagnetic centre, a hyperfine tensor in the effective spin representation is given by $\mathbf{A}^{\mathrm{eff}} = \mathbf{A}\mathbf{g}^{\mathrm{eff}}/g_{\mathrm{e}}$, and the mixing of the

Table 1 Representative range of g-values for selected cytochrome P450s in different stages of the catalytic cycle (see Fig. 1). Effective g-values are given for high-spin species. Simulations using these parameters are given in Fig. 2

P450	Ref.	Intermediate	S	g_x	g_y	g_z
CYP101A1 (P450cam)	[94]	**A** (resting state)	½	2.45	2.26	1.91
P450cam with camphor		**B** (low-spin)[a]	½	2.42	2.24	1.97
		B (high-spin)[a]	$^5/_2$	7.85	3.97	1.78
P450cam with 5-*exo*-hydroxycamphor	[35]	**I** (product bound)	½	2.62	2.25	1.89
CYP3A4	[8]	**A** (resting state)	½	2.421	2.259	1.921
CYP3A4 with 1,2,3-triazole		**B**[b]	½	2.505 2.459	2.259 2.261	1.874 1.890
CYP3A4 with 17-click		**B** (water-bridged, nitrogen of 17-click not coordinated)	½	2.415	2.249	1.922
Characteristic values	[11]	**E** (ferric peroxo)	½	~2.26	2.17	1.96
		F (ferric hydroperoxo)	½	≥2.29	2.18	1.93

[a]At 12 K, a high-spin signal (60%) and low-spin signals from **B** and **A** are observed
[b] two species, possibly corresponding to ligation of the heme through different nitrogen moieties (see Fig. 8c)

$m_s = \pm 3/2$ doublet into the ground-state by the electron Zeeman and electron-nuclear hyperfine interaction requires a tensor representation for a nuclear Zeeman interaction, $\mathbf{g}_n$, which is coaxial with the zero-field splitting tensor [52]. As these spin systems can be treated with an effective S^{eff} spin, the analysis of HYSCORE or ENDOR data can be described using theory from the simpler $S = 1/2$ case with the appropriate scaling for the parameters $\mathbf{g}_n$, $\mathbf{A}$, and $\mathbf{P}$ [51, 52].

The Fe^{II} heme centre can have either an $S = 0$ spin state which is diamagnetic, or an $S = 2$ spin state which is generally referent to as 'EPR-silent' as none of the EPR transitions can be excited at the typical fields and microwave frequencies employed.

To enable electron-nuclear interactions unresolved in CW EPR data to be determined, pulse EPR experiments are employed to record directly a nuclear frequency spectrum. The energy level diagram for an $S = 1/2$, $I = \text{I}$ spin system is given in Fig. 4 and shows the nuclear transitions which hyperfine spectroscopy experiments aim to excite and record (here hyperfine spectroscopy refers to techniques to measure nuclear frequency spectra, i.e. ENDOR, ESEEM, or ELDOR-detected NMR (data from this technique is not given in this review but is mentioned as a technquie gaining importance)).

In particular, the hyperfine interaction of a nucleus provides valuable structural information. The three principal hyperfine values $[A_1, A_2, A_3]$ can be broken up into an isotropic A_{iso} and an anisotropic $[-T_1, -T_2, T_1 + T_2]$ component,

$$\left[A_1, A_2, A_3\right] = A_{iso} + \left[-T_1, -T_2, T_1 + T_2\right]. \quad (5)$$

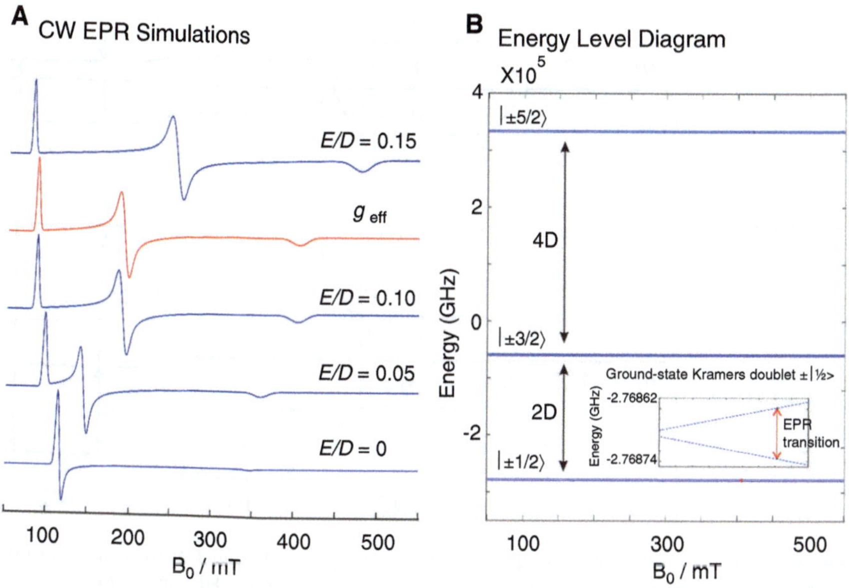

Fig. 3 Simulated X-band (9.7 GHz) EPR data using an effective g-value and the full spin Hamiltonian model for high-spin Fe^{III} heme. (**a**) CW EPR spectra simulated using the full spin Hamiltonian (Eq. (1)) with the indicated *E*/*D* ratios (blue), compared to a simulation using typical effective *g*-values (red) for high-spin Fe^{III} heme (Eq. (4)). The simulation with $E/D = 0.1$ matches closely to the effective *g*-values. (**b**) Energy level diagram for the full spin Hamiltonian, the B_0 field axis is near to the *z*-axis orientation. The diagram shows the three spin states. The $| \pm 1/2 >$ Kramers doublet lies lowest for positive *D* and gives rise to the EPR transitions observed in experiments. Inset: an enlarged view of the splitting of the lowest doublet. Simulation parameters: effective Hamiltonian, $S_{eff} = 1/2$, $g_{eff} = [1.70, 3.62, 8.07]$; full spin Hamiltonian, $S = 5/2$, $g = 2$, $D = 1e8$ MHz ($H_{ZFS} \gg H_{EZ}$)

If the hyperfine coupling is axial, $T_1 = T_2$, allowing the dipolar component to be expressed as $[-T, -T, 2T]$, where T is called the dipolar coupling constant (i.e. Eq. (7)). A_{iso} enables the *s*-type spin density at the nucleus to be calculated. This aids in the characterization of the singularly occupied molecule orbital (SOMO), and the presence of an A_{iso} is useful to establish, for example, that a substrate is electronically connected (thus chemically bound) to the paramagnetic center (of P450), and to characterize spin polarization mechanisms if the sign of A_{iso} can also be determined [53]. The anisotropic component may depend mostly upon the spin density in, for example, the *p*-orbital of nitrogen (e.g. heme ring nitrogens, Fig. 5 shows their nuclear frequencies in a HYSCORE experiment). The conversion from hyperfine coupling to spin population (the integrated spin density of an orbital) can be estimated using computed hyperfine coupling constants from an atomic orbital of the atom and a hybridization model for the atom in the molecule [54], or by comparison to

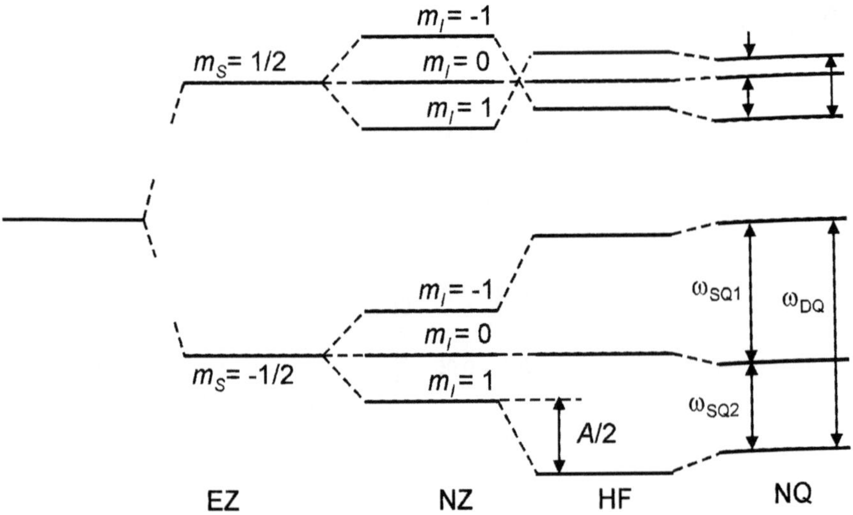

Fig. 4 An energy level diagram for an $S = ½$, $I = 1$ spin system. The nuclear frequencies involve the nuclear Zeeman (NZ) interaction (with known scalar g_n), hyperfine (HF) interaction (tensor **A**) and nuclear quadrupole (NQ) interaction (tensor **P**), see Eq. (1). In the two-dimensional experiment HYSCORE the one double-quantum (DQ) and two single-quantum (SQ1 and SQ2) NMR transitions in each electron spin manifold are correlated against each other (in principle giving 9 possible cross-peaks for each coupled $I = 1$ nucleus)

quantum chemically calculated EPR parameters and spin populations, for example from density functional theory (DFT).

The proton hyperfine coupling is very useful to determine the corresponding electron-nuclear distance as the hyperfine coupling anisotropy is not contributed to by spin density in *p*-type or *d*-type orbitals (for example for a ^{14}N nucleus p-type contributions may need to be accounted for). This allows all the hyperfine anisotropy to be modelled by the electron-nuclear point-dipolar formulae,

$$\mathbf{T} = \frac{\mu_0}{4\pi\hbar} g_e \mu_B g_n \mu_N \sum_i \rho_i \frac{3\vec{n}_i\vec{n}_i^T - 1}{r_i^3} = \sum_i T_i \left(3\vec{n}_i\vec{n}_i^T - 1\right) \qquad (6)$$

where **T** is the dipolar coupling tensor (of size 3 × 3) and $\vec{n}_i$ is a unit vector describing the electron-nuclear direction, with distance r_i, in a molecular frame where vector $\vec{B}_0$ is along the *z*-axis and T_i is given by

$$T_i = \frac{\mu_0}{4\pi\hbar} g_e \mu_B g_n \mu_N \frac{\rho_i}{r_i^3} \qquad (7)$$

Equation (6) is valid for unpaired electron-nuclear separations of *ca.* >2.5 Å and allows a model to be developed where the spin population contributions ρ_i are delocalized. This is implemented in terms of atomic spin populations belonging to

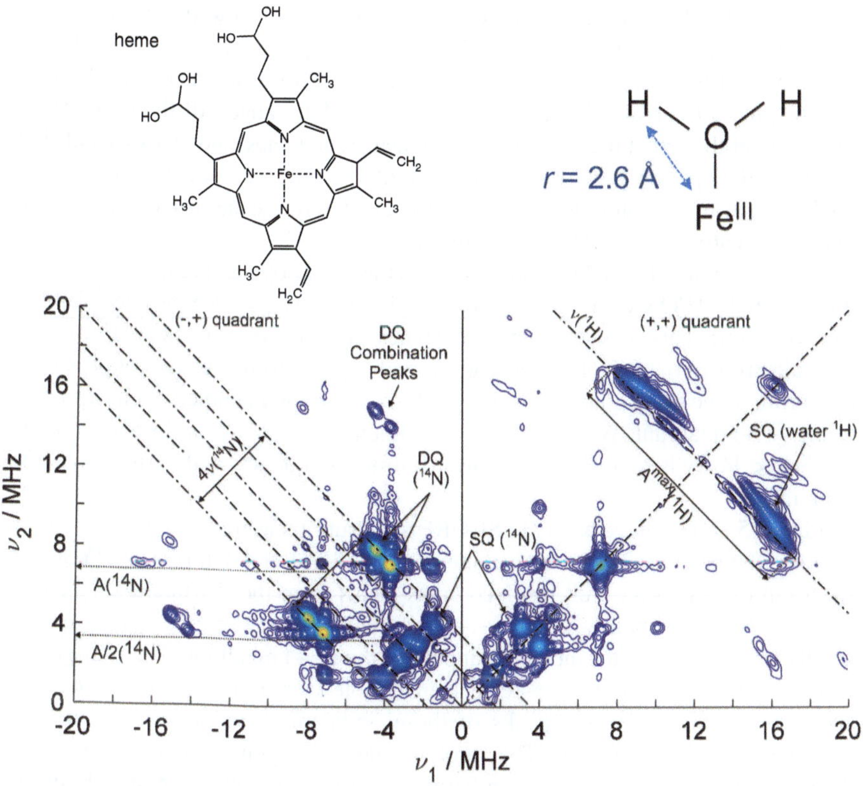

Fig. 5 An X-band HYSCORE spectrum measured at 20 K on the active-site low-spin Fe^{III} heme centre ($S = 1/2$) of P450cin. The spectrum reveals electron-nitrogen and electron-proton couplings. The heme nitrogen cross-peaks are mainly in the (−,+) quadrant and describe two sets of essentially magnetically equivalent ^{14}N ($I = 1$) nuclei. The proton cross-peaks in the (+,+) quadrant from the coordinated water molecule can be used with the point-dipolar formula to determine the proton-to-Fe^{III} (unpaired electron) distance. The dot-dash lines in the (−,+) quadrant are split by $2\nu(^{14}N)$ and $4\nu(^{14}N)$ (ν is the ^{14}N Larmor frequency) and help to identify the ^{14}N cross-peaks in the strong coupling regime. Similarly, the anti-diagonal line at $\nu(^1H)$ identifies ^{1}H cross-peaks in the weak coupling regime. The proton and nitrogen hyperfine couplings, $A(^1H)$ and $A(^{14}N)$, respectively, can be estimated from the peak positons as indicated

nuclei which contribute to the SOMO (e.g. for P450 the Fe^{III} ion and also the directly coordinated heme nitrogens can be included). The point-dipolar approach was used, for example, to characterize structurally the coordination of water to the Fe^{III} ions of P450cam (CYP101A1) (see section "The g-values as an Indicator for Substrate Interactions" [55].

There are three main methodologies used to measure nuclear transitions frequencies when they are not resolved in the CW EPR spectrum [30]: (A) ENDOR (electron nuclear double resonance) which employs excitation using radio frequency pulses in combination with microwave pulses for detection; (B) electron double resonance (ELDOR) detected NMR which requires two mw frequencies one of which is swept; and (C) electron spin echo envelop modulation (ESEEM) techniques which are based on excitation of forbidden EPR transitions in the GHz range by mw pulses to create NMR coherences in the MHz range. All three experiment types aim at the direct measurement of nuclear frequencies, and analysis and simulation of the spectra provides the identity of the nucleus, and its hyperfine and nuclear quadrupole couplings, which are used to determine the structure of the paramagnetic centre as described briefly above.

Of the available ESEEM pulse sequences the most popular and useful is the two-dimensional HYSCORE experiment (HYSCORE = HYperfine Sublevel CORrElation). This pulse sequence correlates the nuclear frequencies from difference electron spin manifolds (see Fig. 4), with signals from weakly coupled nuclei appearing in the (+,+) quadrant, and signals from strongly coupled nuclei appearing in the (−,+) quadrant. Weak coupling is defined by $2|\nu_I| > |A|$, and strong coupling by $2|\nu_I| < |A|$, where ν_I is the nuclear Larmor frequency and A is a hyperfine coupling.

Figure 5 shows an X-band HYSCORE spectrum measured at 20 K on the low-spin Fe^{III} heme centre at the active site of CYP176A1 (P450cin). The HYSCORE spectrum exhibits very well resolved cross-peaks from the ^{14}N nuclei of the heme ring which appear mainly in the (−,+) quadrant, and proton cross-peaks from an axially coordinated water molecule appear in the (+,+) quadrant. The nitrogen signals consist of two sets of cross-peaks from double-quantum transitions ($\Delta m_I = \pm 2$), and also combination cross-peaks of these, indicating two sets of magnetically equivalent nitrogen nuclei. The proton signals depend on the strength of the through-space dipolar coupling between the unpaired electron and ^1H nuclear spin, allowing the electron-nuclear distance of *ca.* 2.6 Å to be readily determined using the electron-nuclear point-dipolar formula, Eq. (6). This example illustrates how hyperfine spectroscopy techniques are able to 'look under' the broad CW EPR linewidth to allow the geometric and electronic structure of the system to be studied in greater detail.

Interpretation of EPR Data Aided by Quantum Chemical Calculations

Quantum chemical calculations using Density Function Theory (DFT) have aided the theoretical understanding of P450 structure and function. The usual approach is to build a computationally manageable model of the active site for the system and to compare DFT computed EPR parameters (e.g. g-values, hyperfine and nuclear quadrupole tensors) with those derived from experiment. This approach helps to validate and refine a proposed chemical model. Ye and Neese composed a thorough review [56], and Hirao et al. reviewed applications to iron-containing molecules of biological interest including P450s [57]. DFT calculations of compound I / compound II models have highlighted features which control reactivity [58].

CW EPR to Characterise the Resting State and Substrate Bound P450 Active Site

This section presents EPR studies characterising resting state ferric P450 and its interaction with water and substrate molecules, i.e. **A** and **B** in Fig. 1.

Extracting Spin-State from g-Values and Its Relation to Optical and MCD Spectroscopy

Based on the observation that the Soret band of the optical spectra of P450s tend to be blue-shifted upon substrate binding, Tsai et al. [21] made early progress (in 1970) concerning the interaction of *Pseudomonas putida's* P450cam and its substrate, D-(+)-camphor. CW EPR was used to characterise the P450cam protein as a typical thio-ligated, hexacoordinate ferric heme centre. At temperatures below 20 K, a high-spin species characteristic of pentacoordinate heme was also observed in small quantities (accounting for <7% of the total CW EPR signal intensity) but, upon substrate binding, this proportion increased to ≈60% of the low temperature CW EPR spectrum. At this early stage, it was not certain whether the thiolate ligand or the axial water ligand was displaced but we now know that these high-spin spectra indicate displacement of the axial water ligand.

Following this early work, Sligar [25] investigated the relationship between optical spectroscopy and the heme Fe^{III} spin-state. Low-spin ferric heme (Fe^{III}, $S = 1/2$) exhibits a Soret peak at $\lambda_{max} = 415–417$ nm while high-spin ferric heme (Fe^{III}, $S = 5/2$) has a Soret maximum at $\lambda_{max} = 390–394$ nm [59]. Optical difference spectroscopy is hence a useful complementary technique to EPR for examining P450 spin state changes when a molecule binds to the active site as shown in Fig. 6.

Because of spectral overlap in the regions of optical spectra relevant to P450 chemistry, difference spectroscopy is conventionally used to characterise spin state changes [59]. Optical difference spectra are characterised as Type I, Type II, or Reverse Type I.

Type I: A molecule binding to P450 that displaces the resting state water without coordinating to the heme (resulting in a pentacoordinate, high-spin ferric heme) causes a blue shift in the heme optical spectrum, resulting in what is known as a Type I optical difference binding spectrum (a maximum at ~390 nm and minimum at ~420 nm) [59].

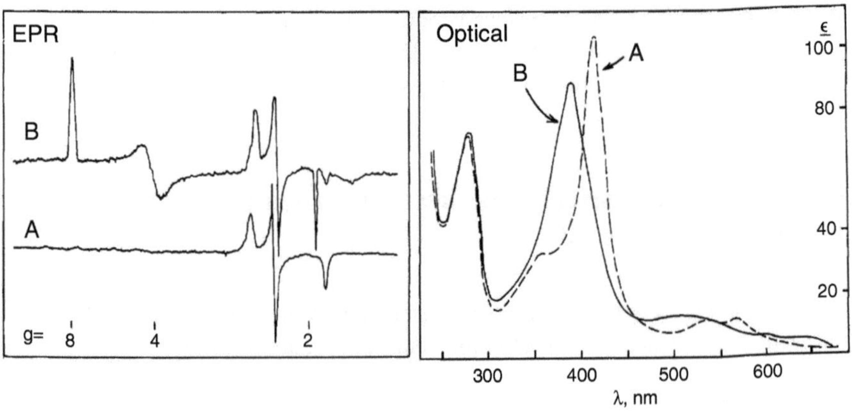

Fig. 6 Comparison of CW EPR spectra and optical spectra from P450cam. Spectra labelled (A) are from P450cam in the resting state (**A**, Fig. 1), displaying a low-spin EPR spectrum and a red-shifted optical spectrum. Spectra labelled (B) are for P450cam with camphor (**B**, Fig. 1), displaying a high-spin EPR spectrum and a blue-shifted optical spectrum. Adapted with permission from [25]. Copyright© (1976) American Chemical Society

Type II: A molecule binding in the P450 active site that displaces the resting state water and then coordinates to the heme centre results in a hexacoordinate, low-spin ferric heme, and optical spectroscopy shows a red shift, resulting in a Type II optical difference binding spectrum (maximum at 425–440 nm, minimum ≤415 nm). Molecules with nitrogen ligands generally bind heme centres strongly and thus Type II spectra have been traditionally associated with nitrogen-ligand inhibitors [59].

Reverse Type I: Some systems start in the high-spin state and transition to the low-spin state upon sample preparation, resulting in an optical difference spectrum that displays the opposite shift to Type I and are characterised by a maximum at 420 nm and a minimum at 390 nm, known as a Reverse Type I spectrum. This is thought to indicate the presence of a weakly-bound endogenous substrate or substrate analogue that is displaced by the target compound [59].

Because of the relationship between the heme spin-state (high-spin and low-spin centres) and the coordination number and type of ligands, optical difference spectroscopy is commonly used to characterise the mode of binding—inhibitor or substrate. However, the spin state of P450s does not always unambiguously reflect the inhibitor or substrate nature of the interaction. For example Connor et al. [8] discovered CYP3A4 exhibits a Type II binding spectrum even though a putative inhibitor molecule does not coordinate to the active site heme as was shown by pulse EPR (section "Interaction of P450 with Substrates in the Resting State"). Based on these data, the authors caution against interpretation of the heme mode of binding solely on the basis of optical spectroscopy [8].

Conner, Atkins et al. [28] investigated the strength of the water coordination for a range of P450s, which is relevant for understanding the differences in substrate binding and spin equilibrium. To this end the differences in the high-spin/low-spin equilibria of various P450 enzymes were characterised using both magnetic circular dichroism (MCD) and g-values determined by CW EPR. The g-values were interpreted using Taylor theory [27] to characterise axial (Δ/λ) and rhombic (V/λ) crystal field terms,

$$V/\lambda = g_x\left(g_z + g_y\right) + g_y\left(g_z - g_x\right) \tag{8}$$

$$\Delta/\lambda = g_x\left(g_z + g_y\right) + g_y\left(g_z - g_x\right) - \tfrac{1}{2}\left(V/\lambda\right), \tag{9}$$

where λ is the spin-orbit coupling constant for Fe^{III}. MCD data from the corresponding set of P450s in the water bound resting state were used to investigate the spin state equilibrium on ligand field strengths. Analysis of the data revealed a considerable difference in the H_2O axial ligand field strengths between different low-spin P450s. These energetic differences will make a contribution to the substrate-dependent spin state perturbation. However, the P450 spin equilibrium in the resting state was found not to be proportional to the low-spin H_2O axial ligand strength, suggesting the spin state equilibrium is dependent upon water accessibility to the heme centre and/or differences in the substrate-free, high-spin state ligand field energies. When comparing MCD parameters with those calculated from EPR measurements using Taylor theory, Connor and co-workers showed that EPR data alone is insufficient for a reliable and accurate description of the MCD transitions [27, 28].

The g-Values as an Indicator for Substrate Interactions

Human aromatase (CYP19A1), a microsomal P450 involved in converting androgens to oestrogens, provides an example where X-band CW EPR spectroscopy has been used to characterise the spin state of P450s in the resting state and in the presence of substrates and inhibitors [22, 23]. Gantt, Sligar et al. [23] investigated the free and substrate bound resting states of aromatase using X-band CW EPR spectra measurements at 20 K. The substrate-free state exhibits two low-spin CW EPR species, which upon binding of the substrate androstenedione are converted to a spectrum comprising two low-spin species and a single high-spin species. These data show that the resting state of CYP19A1 under these conditions exhibits conformation heterogeneity. 19-Hydroxyandrostenedione-bound aromatase only exhibited one low-spin species, allowing the authors to easily distinguish between bound and unbound aromatase from the conversion of the CW EPR spectrum with two species to one with only one species. Notably, the only observed difference between the principle *g*-values for 19-hydroxyandrostenedione-bound and substrate free CYP19A1 was in the high-field *g*-value. Gantt and Sligar's cryoreduction work to characterise aromatase reaction intermediates [23] will be described below (see section "High-resolution Pulsed EPR to Characterise the Resting State and Substrate Bound P450 Active Site").

Maurelli et al. [24] investigated non-steroidal inhibitors of aromatase which are relevant for the treatment of oestrogen-dependent tumours such as those found in breast cancer. They compared the low-spin *g*-values for substrate free aromatase with those from the enzyme bound with the non-steroidal inhibitor anastrozole. They observed changes only in the high-field principle *g*-value upon addition of anastrozole (as for 19-hydroxyandrostenedione-bound aromatase). The authors observed that the two-species, substrate-free aromatase CW EPR spectrum was converted into a single-species spectrum upon binding of anastrozole, suggesting that the enzyme exists in multiple conformations which are converted to only one conformation upon substrate binding. This, along with the Type II optical binding spectrum motivated the use of higher resolution pulsed EPR techniques to further characterise the interaction of substrates with the aromatase active site (section "Interaction of P450 with Substrates in the Resting State").

It has been observed for some substrate-bound P450s that binding of a redox partner has a significant effect on the CW EPR spectrum, as reported by Takeuchi et al. [29] for P450cam (redox partner putidaredoxin, Pdx) and CYP11A1 (P450scc, a cholesterol side-chain cleavage catalyst in mammals; redox partner adrenodoxin, Adx). Addition of the redox partner to the resting state of either P450cam or P450scc showed no detectable change in the CW EPR spectrum, but addition of Pdx to substrate-bound P450cam resulted in a shift of the spin-state equilibrium towards low-spin, accompanied by significant changes in both the high- and low-spin *g*-values. For P450scc with cholesterol bound, the spin equilibrium instead shifted towards high-spin, and while a significant change was seen in the high-spin *g*-values, the low-spin *g*-values remained unperturbed. This suggests perturbation of the heme plane in this high-spin species, as the zero-field splitting is very sensitive to the geometry of the heme plane. For P450scc bound to its second stage substrate, 22*R*–hydroxycholesterol, neither a low-spin *g*-value shift nor a spin equilibrium shift was observed (P450scc + 22*R*–hydroxycholesterol is entirely low-spin with and without Adx). To further investigate the interaction between the P450 and redox partner protein, Takeuchi and co-workers also measured the spectrum of the iron-sulphur clusters of the redox partner when bound to a reduced (and EPR-silent) P450 but found that the EPR spectrum of neither Pdx nor Adx changed significantly upon addition of P450 or substrates; the only significant shift was for Pdx + P450cam + substrate in the presence of CO.

Johnston et al. [7] utilised CW EPR spectroscopy to show that P450cin (CYP176A1) and P450cam exist as both Fe^{III} and Fe^{II} in the resting state in *Escherichia coli* (a number of mammalian P450s were also characterised using Fe^{II}-CO versus Fe^{II} difference spectroscopy). Because *E. coli* lacks native P450s, endogenous reduction of P450cam and P450cin could be studied in the absence of their electron transport chain partners (putidaredoxin and putidaredoxin reductase for P450cam, and cindoxin for P450cin). The authors observed an increase in the intensity of the Fe^{III} heme CW EPR signal after exposing the cells to O_2 to oxidise the P450s, suggesting that significant amounts of resting state P450 are endogenously reduced to an EPR-silent Fe^{II}-P450 state, contrary to the common depiction of only a Fe^{III}-P450 resting state (Fig. 1—resting state **A**) of the P450 catalytic cycle *ex vivo*.

In summary, when hexacoordinated the Fe^{III} heme is in a low-spin state and is high-spin when pentacoordinated. In the P450 resting state, the heme is typically coordinated by water (distal ligand) resulting in a low-spin CW EPR spectrum. This spectrum is often comprised of several species, presumably the result of small differences in the binding interactions of the coordinated water. Binding of substrate to the P450 active site but not directly to the heme typically displaces the water giving a pentacoordinate high-spin complex. It is, however, possible to bind substrate without displacing the distal water ligand and this situation can be detected by a shift of the low-spin resting state *g*-values. If a target molecule coordinates the heme, then a low-spin EPR spectrum is observed which generally displays characteristic *g*-values. Many P450s display a mixture of high-spin and low-spin species and the percentage can be characterised at room temperature by optical difference spectroscopy or at low temperature by EPR spectroscopy. Often the percentage of low-spin species is higher in frozen solution (as determined by CW EPR), indicating a repositioning of the substrate in the heme pocket which allows water to re-bind.

High-Resolution Pulsed EPR to Characterise the Resting State and Substrate Bound P450 Active Site

In combination with information from CW EPR (as described above), pulse EPR has been successful in structurally characterising the interaction of resting state ferric P450 enzyme with water and substrate molecules, **A** and **B** in Fig. 1. A selection of interesting examples follows in which it was possible to determine the axial ligands to the Fe^{III} heme. The review by Cruce [60] provides further examples of pulse EPR studies on P450 enzymes.

Nitrogen Ligands in Heme Model Complexes and Cytochromes

Although not a cytochrome P450, the study of cytochrome *b*559 (a heme protein constituent of photosystem II in spinach) by García-Rubio et al. [61] using HYSCORE spectroscopy proved instructive in particular for the identification and interpretation of characteristic nitrogen peaks from the heme. To distinguish between the porphyrin nitrogens and two nearby histidine residues of cytochrome *b*559, the authors compared the HYSCORE spectrum for cytochrome *b*559 with those of two bis-imidazole heme complexes. By isotopically labelling the model system using a combination of ^{14}N and ^{15}N nuclei, the nitrogen nuclear quadrupole

(NQ) and hyperfine couplings of the imidazole and pyrrole moieties were determined. The hyperfine couplings indicate that the unpaired electron is confined mainly to a non-bonding iron orbital, and the NQ axes enabled the orientation of the axial ligands to be determined. The data and analysis characterise the electronic properties of the heme centre, and were discussed in terms of the electron transfer function of the protein.

Resting State of P450 with Coordinated Water

Electron spin echo envelope modulation (ESEEM) techniques measure a nuclear frequency spectrum by exciting nuclear transitions solely with microwave frequency radiation. This approach has been utilised to investigate the binding of water to the Fe^{III} heme centre of P450cam using ^{17}O water (^{17}O: $I = 5/2$) [62]. The division of two-pulse ESEEM traces (one from an ^{17}O–enriched water sample and the other from an unenriched ^{16}O ($I = 0$) water sample) and subsequent Fourier transformation enables a nuclear frequencies spectrum comprised of only ^{17}O contributions to be obtained. Numerical simulations were then matched to the experimental nuclear frequency spectra to determine the ^{17}O hyperfine and nuclear quadrupole couplings. These data led the authors to conclude that the distal ligand of resting state P450cam is a water molecule. Analysis of the orientation of the ^{17}O nuclear quadrupole tensor additionally allowed determination of the orientation of the water molecule, revealing that the Fe-O axis was 18 ± 3° from the heme ring normal (Fig. 7).

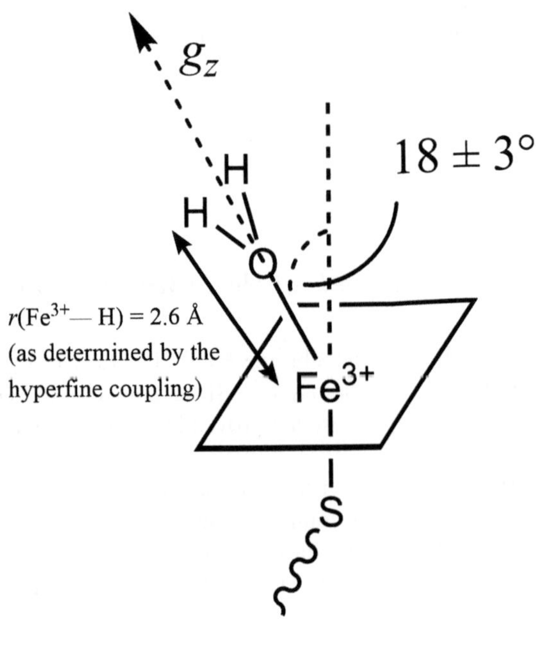

Fig. 7 Thomann et al. [62] utilised ESEEM and ENDOR measurements on P450cam samples in the resting state, prepared with $H_2{}^{17}O$ or ^1H_2O, to identify that the proximal ligand is water. The angle between the Fe^{III}—O bond and the axis normal to the heme is 18 ± 3°, and the distance $r(Fe^{III}—H) = 2.6$ Å

Goldfarb et al. subsequently employed a combination of three-pulse ESEEM and pulse ENDOR to investigate the protons of the coordinated water [55]. The data provided the ^{1}H hyperfine tensor (principal values and axis orientations) to be determined which enabled an estimation of the proton-electron distance of 2.6 Å using the point-dipolar formulae (Eq. (6)). Combined, the ^{17}O and ^{1}H EPR data were interpreted in terms of an axially bound water molecule with two equivalent protons. The data and analysis also eliminated the possibility of the resting state axial ligand being an OH^{-} molecule.

Interaction of P450 with Substrates in the Resting State

Conner et al. [8] investigated the interaction of CYP3A4 with the inhibitor 1,2,3-triazole (1,2,3-TRZ), the inactivator 17-α-ethynyloestradiol (17EE), and the so-called 17-click which incorporates a 1,2,3-TRZ fragment into 17EE, (Fig. 8b, c, d

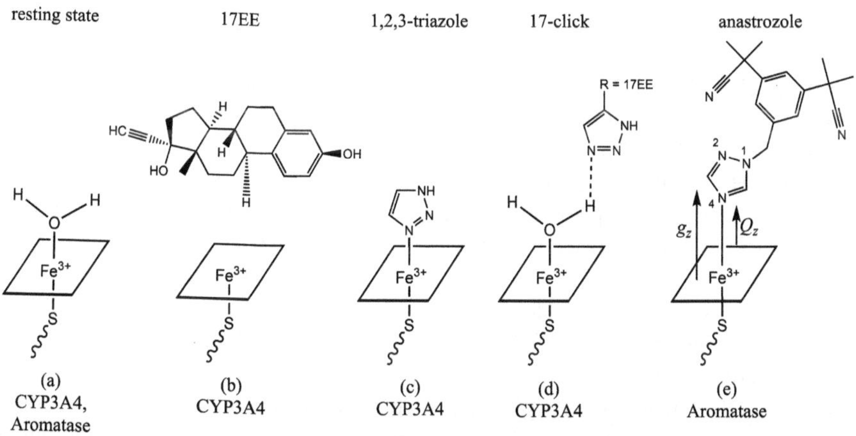

Fig. 8 Examples of active site interaction classes in P450s. (**a**) The resting state (Fig. 1, **A**) is a water-ligated, hexacoordinated, low spin $S = 1/2$ Fe^{III} heme (there is some evidence that the resting state *in vivo* can also be in the Fe^{II} state [7]). The HYSCORE spectrum shown in Fig. 9 (*left*) is observed, exhibiting distinctive water proton peaks. (**b**) In a typical P450, the substrate binds to P450 but does not ligate the heme centre, resulting in a pentacoordinated, high-spin $S = 5/2$ species, corresponding to **B** in Fig. 1. In this case a Type I optical absorption difference spectrum is observed. Shown here is 17-α-ethynyloestradiol (17EE) [8]. (**c**) Typical interaction of a nitrogen ligand which coordinates to the Fe^{III} heme, inhibiting the enzyme. In this case a Type II optical absorption difference spectrum is observed. Shown here is 1,2,3-triazole [8]. The HYSCORE spectrum of (**c**) is shown in Fig. 9 (*centre*) and lacks the distinctive water proton peaks of Fig. 9 (*left*). (**d**) Water-bridged inhibitor as is observed in CYP3A4 [8]. The substrate comprises 1,2,3-triazole incorporated with 17EE (17-click). Although HYSCORE spectroscopy clearly shows the presence of a coordinated water molecule (Fig. 9 (*right*)), a Type II difference optical absorption spectrum is observed which is typical of inhibitor binding. (**e**) Aromatase with bound inhibitor anastrozole, with N4 assigned using information from the measured hyperfine couplings. The nuclear quadrupole tensor z-axis and g factor z axis directions are approximately parallel as determined from analysis of the quadrupole coupling tensor (discussed in [24])

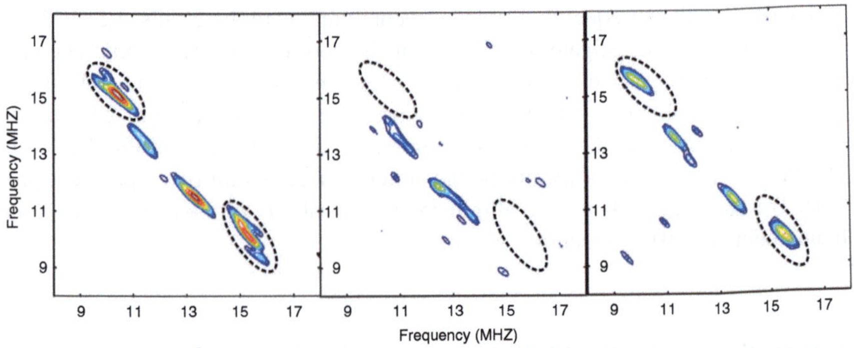

Fig. 9 X-band HYSCORE spectra of (*left*) substrate-free CYP3A4 (Fig. 8a), showing distinctive water proton hyperfine couplings (*circled*); (*centre*) CYP3A4 + 1,2,3-triazole (Fig. 8c) lacking the distinctive water proton hyperfine couplings (*circled*); and (*right*) CYP3A4 + 17-click (Fig. 8d) showing distinctive hyperfine couplings from water protons (*circled*), indicating that a water molecule remains coordinated to the Fe^{III} heme centre despite binding of 17-click to the protein. The CYP3A4 sample concentration was approximately 50 μM, and the data were measured near to the field position of the low-field principle *g*-value of the heme spectrum. Figure adapted with permission from [8]. Copyright© (2012) American Chemical Society

respectively). Both 1,2,3-TRZ and 17-click molecules contain a nitrogen atom in the 1,2,3-TRZ fragment for potential coordination to the Fe^{III} heme. The interaction of these molecules with the active site of CYP3A4 was probed with optical, CW EPR and HYSCORE spectroscopy. CYP3A4 + 17EE (Fig. 8b) displayed a high-spin EPR spectrum (type I optical). CYP3A4 incubated with either 1,2,3-TRZ (Fig. 8c and Fig. 9 (centre)) or 17-click (Fig. 8d and Fig. 9 (right)) displayed a typical low-spin (type II optical) spectrum, suggesting both act as inhibitors by coordinating via a nitrogen atom. However, the CW EPR spectra revealed distinctively different *g*-values between CYP3A4 + 1,2,3-TRZ and CYP3A4 + 17-click, suggesting a different coordination mode between the samples. The structure was revealed by HYSCORE: whilst CYP3A4 + 1,2,3-TRZ lacked proton signals from coordinated water (inferring that the axial water ligand had been replaced by the inhibitor), the typical proton signals from coordinated water were clearly still visible for CYP3A4 + 17-click, despite the type II optical spectrum. The structural information from HYSCORE demonstrates (a) that some P450 enzymes can form a water-bridged, low-spin heme complex with a nitrogen ligand, and (b) that optical spectroscopy cannot be used as the sole basis for identifying the nature of the ligand interactions with P450.

As mentioned earlier, Maurelli et al. [24] utilised CW EPR and HYSCORE to show and accurately characterise the coordination chemistry of the non-steroidal inhibitor anastrozole to the Fe^{III} heme of aromatase (CYP19A1), Fig. 8e. HYSCORE before and after addition of the inhibitor enabled nitrogen (^{14}N) nuclear frequencies from the heme and the anastrozole ligand to be confidently assigned. Simulations of the anastrozole signals provided the full hyperfine (HF) and nuclear quadrupole (NQ) tensors of a strongly coupled ^{14}N nucleus. The HF coupling is consistent with coordination of anastrozole via N-4, Fig. 8e. This was reinforced by the size of the NQ coupling and the direction of the unique principal axis which is along the nitrogen lone pair donor orbital and collinear with the g_z axis which is normal to the heme plan. The CW EPR and HYSCORE data thus showed anastrozole to be coordinated to Fe^{III} via a σ-bond formed from the overlap between the lone pair of the imidazole nitrogen N-4 and the empty iron dz^2 orbital.

Roberts et al. [63] used HYSCORE techniques to investigate CYP2C9, a human cytochrome P450 critical to xenobiotic metabolism and capable of accessing a very wide range of substrates. Protein conformational flexibility was investigated with a R108H mutant which showed that the histidine of this mutant was able to move about 12 Å from its predicted position, based on a CYP2C9 crystal structure, to coordinate to the Fe^{III} heme ion. The determined coordination in the R108H mutant was based upon a number of experimental measurements. UV-vis absorbance measurements yielded a spectrum characteristic of P450 coordination by a nitrogen ligand. Aggregation studies showed that this coordination was not the result of protein oligomers, i.e. intermolecular protein interactions could be by ruled out with size exclusion chromatography. Finally, X-band HYSCORE experiments were used to definitively determine the identity of the ligand. EPR data were recorded from a wild-type (WT) CYP2C9, a WT CYP2C9 + imidazole, and a R108H mutant sample. Both the CYP2C9 + imidazole and R108H mutant showed that signature ^1H HYSCORE signals from axial coordinated water were absent, and that ^{14}N HYSCORE cross-peaks from the heme were essentially identical, and different from WT CYP2C9 without the imidazole. The data allowed the authors to conclude the bound ligand was the imidazole moiety of a histidine residue. They were also able to examine the flexibility of the protein in terms of open and closed structures and to posit explanations for P450 substrate-promiscuity.

Characterisation of the P450 Reaction Cycle by EPR

Characterisation of the P450 catalytic cycle using EPR has focused on methods to isolate and trap the short-lived intermediates shown in Fig. 1 (intermediate **D–I**). Cryoreduction and/or rapid freeze-quench has enabled the following chemical intermediates to be trapped and unambiguously identified through a combination of CW and pulse EPR studies [9–12, 31–35]: **E** (ferric-peroxo, Fe^{III}-O_2^-); **F** (ferric-hydroperoxo, Fe^{III}-OOH); **G** (iron(IV) oxo (ferryl) porphyrin π-cation radical, $Fe^{4+} = O\ P^{\bullet+}$); **I** (product-bound intermediate); and potentially even an iron(V) intermediate for metal complex analogues of P450 systems [64]. Intermediate **C** is EPR-silent, and **D** and **H** diamagnetic, and have been characterised in detail by other techniques including Mössbauer, X-ray absorption and UV-VIS spectroscopy [65]. To aid in interpretation, the EPR data from P450s have been extensively compared with related oxidising hemoproteins (e.g. nitric oxide synthase (NOS) [66, 67], chloroperoxidase (CPO) CPO: [38, 65, 68], peroxygenase (APO) [69, 70], heme oxygenase (HO) [71, 72], and *Amphitrite ornate* dehaloperoxidase [33]). Evidence from both P450s and related systems have suggested species **E**, **F** and **G** form a series of increasing chemical reactivity, with the intermediate that is catalytically active depending primarily on the substrate molecule (and to a lesser extent on other accessory molecules and the active site configuration [73]). Particularly valuable has been investigation of Compound I in model systems [11, 33]. This section describes firstly the generation and characterisation of compound I, followed by a description of the information gleaned by cryoreduction experiments.

Characterisation of Compound I and Its Reactivity

Compound I is generally accepted to be the active species for the majority of P450 monooxygenation reactions, and of all monooxygenation insertions into C–H bonds. For more than 40 years spectroscopic data on Compound I in a P450 enzyme remained inconclusive [19] until Green's seminal study on CYP119A1 [13]. This work showed how to isolate compound I in high yield, and presented a detailed characterisation of its electronic structure using EPR, Mössbauer, and X-ray absorption spectroscopies. It is interesting, if not remarkable, that the most important factor enabling this impressive feat proved to be a high purity sample free from unwanted substrates bound at the active site, as described in the minireview by Green [19]. The unwanted substrates adversely change the rates of compound I formation and decay by approximately ten fold.

Using high purity ferric CYP119 mixed with *m*-chloroperbenzoic acid for 35 ms enabled Green and co-workers to obtain compound I with a yield of 75% as determined by stop-flow experiments combined with UV-vis spectroscopy. Samples of CYP119 containing high-yields of compound I for spectroscopic characterization were then obtained using rapid freeze-quench experiments. In their protocol for this enzyme, ferric CYP119 (3 mM) is mixed (2:1) with a 12 mM solution of *m*-chloroperbenzoic acid, and the aqueous reaction mixture sprayed into liquid ethane (89 K) 3.5 ms after mixing.

The CW EPR spectrum of compound I (see Fig. 10) is interpreted in terms of an $S_1 = 1$ iron(IV)oxo unit (with principal values for CYP119 of $g = 2.27, 2.20, 2.00$) exchange coupled with an $S_2 = 1/2$ ligand-based radical ($g' \approx 2$) [68]. The relevant spin Hamiltonian is given by

$$H_0 = \mathbf{D}\left(\hat{S}_{1Z}^2 - \frac{2}{3}\right) + J\vec{S}_1^T\vec{S}_2 + \beta\vec{B}_0^T g^{ferryl}\vec{S}_1 + \beta g'\vec{B}_0^T\vec{S}_2 \qquad (10)$$

Doublet and quartet spin states result from the exchange coupling J, which are mixed by the strong zero-field splitting D of the ferryl moiety, and result in a spin system with three-Kramers doublets where only the lowest is populated at the temperature of the EPR experiments. This description allows an effective spin Hamiltonian model to be used that is described by three effective g-values, obtained by simulation of the spectrum from CYP119, $g_{eff} = 1.96, 1.86, 2.00$. The values are indicative of antiferromagnetic coupling and a $|J|/D = 1.30$. Figure 11 provides an

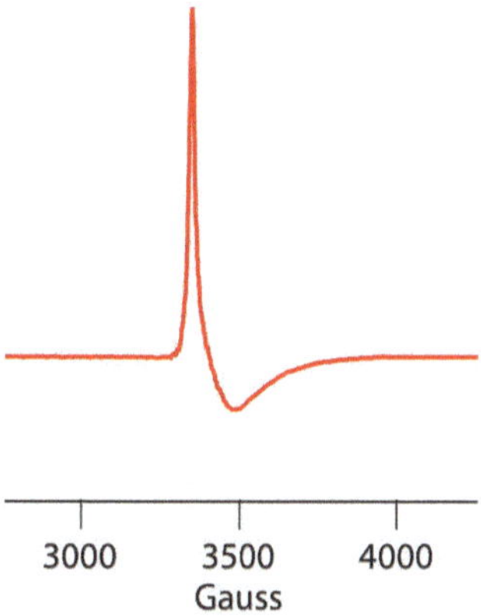

Fig. 10 X-band CW EPR spectrum (recorded at 7.5 K) of Compound I in CYP119A1 prepared using the peroxide shunt method. The spectrum was produced by subtracting data measured with a microwave power of 128 mW from data recorded at 203 mW to remove unwanted paramagnetic signals. Adapted from [14]. Adapted with permission from AAAS. For the electronic imprints of this book, readers may view, browse, and/or download this figure for temporary copying purposes only, provided these uses are for non-commercial personal purposes. Except as provided by law, this material may not be further reproduced, distributed, transmitted, modified, adapted, performed, displayed, published, or sold in whole or in part, without prior written permission from the publisher

overview of the EPR properties of this spin-state. In their paper, Green and co-workers also presented data showing that compound I of CYP119 is catalytically competent, strengthening the case for it being the catalytically active intermediate of all P450 monooxygenation reactions [13].

Following the success with CYP119A1, the same preparative methods were employed to isolate and characterise compound I in $P450_{ST}$ (CYP119A2), P450cam and CYP158 (Table 2). A collection of compound I spin Hamiltonian parameters are shown in Table 2 for P450 and CPO. The high-yield generation of compound I has allowed detailed spectroscopic data from a range of techniques [13, 14, 19, 38, 65, 74] to be collected that has advanced the understanding of P450 reactivity enormously [36]. Here we compare the reactivity of P450 to CPO (a thiolate-ligated heme enzyme), which is less reactive than P450 and is only capable of cleaving activated C–H bonds. The EPR spectrum of CPO compound I has similar features to that of P450 compound I and can be simulated with g_{eff} = 1.72 1.61 2.00, and $|J|/D$ = 1.02 (Table 2). The $|J|/D$ ratio is 30% lower than in P450 (Table 2). To obtain values for $|J|$ and D (rather than the ratio determined by EPR), variable temperature Mössbauer spectroscopy experiments were carried out. These established that the zero-field splitting interaction D is very similar between the two enzymes, and thus the difference in the $|J|/D$ ratio indicates a larger exchange coupling J in P450 (Table 2) [38]. Further characterisation with EXAFS spectroscopy determined the Fe–S bond to be shorter in P450 compared to CPO (2.388 Å compared to 2.481 Å). The Fe–O bond distances are comparatively longer in P450 than CPO (1.670 Å

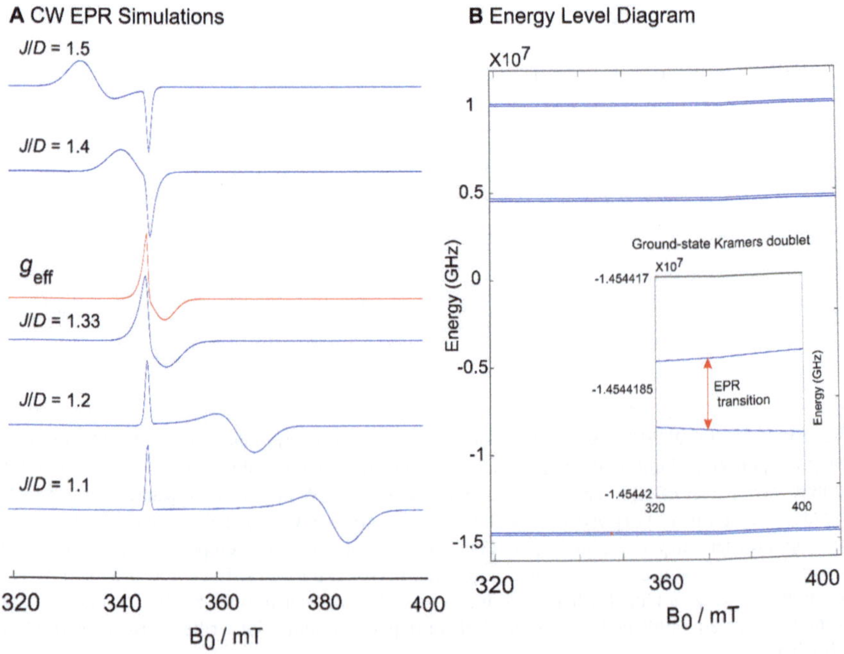

Fig. 11 Simulated X-band (9.7 GHz) EPR data showing the method of analysis of the compound I paramagnetic centre (note these simulation use axial g-values). (**a**) CW EPR spectra simulated using the full spin Hamiltonian (Eq. (10)) with the indicated J/D ratios (*blue*), compared to the one using effective g-values (*red*) which describe the experimental data [13]. The simulation with J/D = 1.33 matches closely the simulation with the effective g-values. (**b**) Energy level diagram for the full spin Hamiltonian, the B_0 field axis is near to the z-axis orientation. The diagram shows the three spin states. The $|\pm 1/2>$ Kramers doublet lies lowest for antiferromagnetic coupling and gives rise to the EPR transitions observed in experiments. Inset: an enlarged view of the splitting of the lowest doublet. Simulation parameters: effective spin Hamiltonian, S = 1/2, g_{eff} = [1.986, 1.986, 2.00]; full spin Hamiltonian: Spin 1: $S_1 = 1$, g_1 = [2.27, 2.27, 2], D_1 = 1e10 MHz ($H_{ZFS} \gg H_{EZ}$). Spin 2: S_2 = 1/2, g_2 = 2. The values for J/D are indicated on the graph

Table 2 Representative range of Spin Hamiltonian parameters for compound I in selected systems

		S = 1/2 representation			(MHz)			Ferryl values		
System	Ref.	g_x	g_y	g_z	D	$\|J\|$	$\|J/D\|$	g_x	g_y	g_z
CYP119A1	[13]	1.96	1.86	2.00	1.20×10^6	1.56×10^6	1.30	2.27	2.20	2.00
$P450_{ST}$ (CYP119A2)	[19]	1.95	1.84	1.99	–	–	1.3	–	–	–
P450cam	[14]	1.98	1.93	2.00	–	–	1.38	2.24	2.21	2.00
CYP158	[65]	2.02	1.99	1.93	–	–	–	–	–	–
CPO	[13]	1.72	1.61	2.00	1.17×10^6	1.20×10^6	1.02	2.27	2.19	2.00

versus 1.661 Å). The Fe–S bond distances appear to be correlated to the $|J|$ values (though there is limited data). Combined the results are interpreted as showing an increase in electron donation from the Cys-S ligand into the ferryl π^* orbitals in P450, as compared to CPO. Importantly, it is known from work on synthetic model complexes that donating axial ligands enhance reactivity with respect to oxygen transfer and hydrogen abstraction. The properties of compound II (**H**) are also key in developing the reactivity model. The Fe–S bond of P450 compound II is shorter than that of CPO (by 0.12 Å) while the Fe–OH bond is longer (by 0.02 Å) [65, 75]. Calculations have shown that the species which are most reactive towards C–H bonds have electron donating ligands and thus the longest Fe–OH bonds. This provides a rationale for the reactivity difference between P450 and CPO enzymes. The differences in the electron donation and therefore the bond lengths and strengths between P450 and CPO, both of which have the same heme-thiolate centre, is reported to be controlled by differences in hydrogen bonding to the axial thiolate within the 'cys-pocket' [38].

From these studies, key factors explaining P450 reactivity have emerged. The basicity of the ferryl oxygen is a key parameter and is characterised by the pK_a (acid dissociation constant) for the equilibrium between compound I and compound II,

$$\text{Cys-S-Fe}^{\text{IV}} = \text{O P}^{+\bullet} \underset{-\text{H}}{\overset{+\text{H}}{\rightleftharpoons}} \text{Cys-S-Fe}^{\text{IV}}\text{-OH P.}$$

The strong basicity of the ferryl oxygen is a result of the strong electron 'push' from the axial thiolate ligand, which leads to a greater 'pull' from the ferryl oxygen for C–H bond cleavage [37] (see the rebound mechanism in Fig. 1). Green and coworkers determined a pK_a of 12 for compound II in two P450s [65] (a value similar to APO, a heme-thiolate peroxygenase with similar reactive to P450, where pK_a = 10 [70]). The K_a for compound II P450 is about 9 orders of magnitude more basic than compound II in histidine-ligated hemoproteins, where p$K_a \approx 3.5$ [65, 76]. This data emphasises the role of the strong electron donor thiolate ligand which increases the basicity of the Fe^{IV}-O moiety and lowers the compound I redox potential. The radical rebound mechanism (Fig. 1) features hydrogen abstraction by the Fe^{IV}=O unit of compound I, and it has been shown that the driving force for C-H activation ΔG_p is related to the energy difference between the O–H bond formed, D(O–H), and the C–H bond broken, D(C–H); ΔG_p = D(C–H)–D(O–H) [77]. For heme enzymes, D(O–H) is given by

$$\text{D}(\text{O}-\text{H}) = 23.06E^0_{\text{compoundI}} + \text{p}K_{\text{a,compoundII}} + C\left(\text{kcal.mol}^{-1}\right) \tag{11}$$

where $E^0_{\text{compound I}}$ is the one electron reduction potential and C is a constant that depends upon solvent and reference electrode [77]. The available P450 data indicate that the high pK_a of compound II offsets the compound I redox potential which is lowered by the thiol ligation. These competing factors allows P450 monooxygenases to oxidise a C–H bond while suppressing long range electron transfer rates, and thus unwanted oxidation of the protein scaffold, that would be a consequence of a higher redox potential according to Marcus theory [65]. This is in contrast to related enzymes with histidine ligation such as peroxidases [37].

Cryoreduction Experiments Combined with EPR Spectroscopy

The development and use of cryoreduction experiments has proven very fruitful in P450 research [11] . Firstly, samples are prepared in the intermediate **D** state by chemical reduction of the P450 resting state, for example with sodium dithionite (**B** → **C**), followed by oxygenation of the sample (in e.g. ethylene glycol) at typically 200–260 K (**C** → **D**). The second reduction (Fig. 1, **D** → **E**) occurs under cryogenic conditions by bombarding the sample with ionising γ-radiation to cause radiolytic cleavage of bonds somewhere in the protein or in nearby water molecules [3], with a liberated electron able to migrate to the active site as a position of low potential energy. Mobile electrons can also be introduced, for example, from high-energy electron sources.

Irradiation of intermediate **D** at liquid nitrogen temperatures, $T \approx 77$ K, may result in trapping of either a ferric peroxy species (Fe^{III}—O_2^-), or a ferric hydroperoxy species (Fe^{III}—OOH), which can be assigned to the catalytic intermediates **E** and **F**, respectively (the basis for these assignments is described below). For example, cryoreduction of P450cam (from intermediate **D**) leads to a ferric hydroperoxy species at 77 K, or a ferric peroxy species at $T < 55$ K. The temperature dependence is a result of the delivery of a proton to the heme centre at temperatures above *ca.* 55 K. Proton transfer can be disrupted by mutating residues comprising the transfer chain e.g. threonine 252 to alanine or aspartate 251 to asparagine in P450cam, which allows the ferric peroxy species to be exclusively trapped in cryoreduction experiments carried out at 77 K. The trapped peroxy or hydroperoxy species, and subsequent species formed after annealing the sample at elevated temperatures for short time periods, can then be studied by spectroscopic techniques, e.g. CW and pulse EPR, Mössbauer spectroscopy, etc. [3, 11].

As described above, cryoreduction generates and traps, from intermediate **D**, one electron reduced oxyheme complexes. Are these cryoreduction species intermediates of the catalytic reaction preformed under physiological conditions? Very significantly, annealing experiments of the cryoreduced samples at higher temperature have shown that cryoreduction species are catalytically competent and lead to the expected product of a particular reaction [11, 31, 34, 35]. Pulse EPR data have enabled the structure of key cryoreduced species to be determined, for example ferric peroxy (Fe^{III}—O_2^-) and hydroperoxy (Fe^{III}—OOH) species. Combined the kinetic and spectroscopic data have enabled key cryoreduced species to be assigned to intermediates of the catalytic cycle carried out under physiological conditions. Henceforth in this review we will directly assign cryoreduction EPR species to intermediates of the catalytic cycle (Fig. 1) where appropriate, and leave it to the interested reader to consult the original literature were the basis for the assignments are covered in detail.

Intermediates **E** and **F** are low-spin $S = 1/2$ species and exhibit distinctive principal *g*-values (**E**: ~2.26, 2.17, and 1.96, **F**: ≥2.29, 2.18, and 1.93, see Table 1 and Fig. 2) [11] which are relatively insensitive to differences between P450s and can thus be used as a fingerprint to identify their presence. Figure 12 shows CW EPR

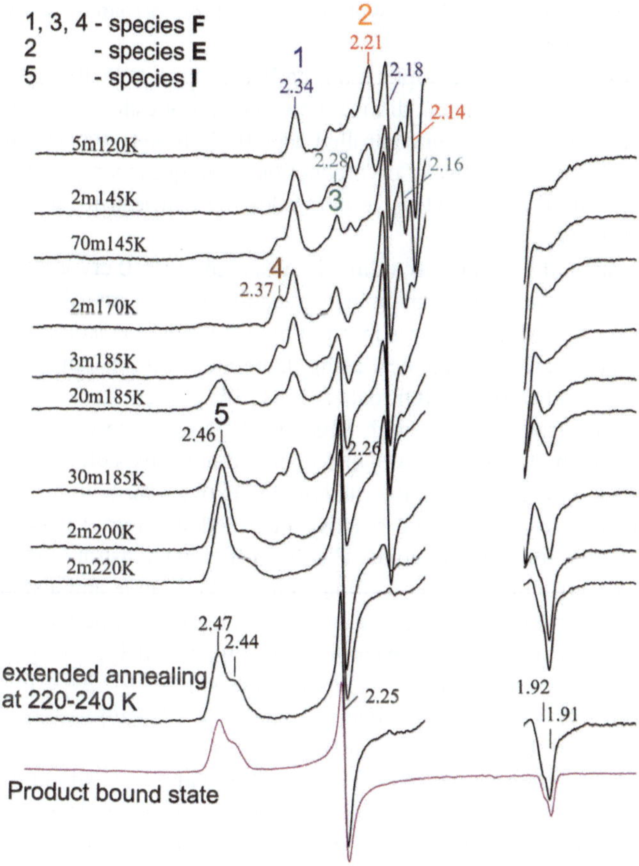

Fig. 12 EPR spectra (in *black*) used to characterise the reaction of cholesterol with oxy-P450scc. The enzyme is firstly cryoreduced to prepare intermediate **E** which is subsequently annealed for various lengths of time at various temperatures (shown in minutes and Kelvin for each spectrum). The bottom spectra were recorded after extended annealing at 220–240 K (*black*) and in *purple*, Fe^{III} P450scc with 22-hydroxycholesterol added (the hypothesised reaction product). On the basis of *g*-factor comparison, the signals from 1 and 3 are assigned to hydroperoxy ferric P450 species (**F**), and signal 2 to peroxy ferric P450 species (**E**). Observation of the hydroperoxy intermediate **F** before annealing indicates a proton transfer mechanism operates at 77 K (as is seen in P450cam and heme oxygenase, with an ordered water molecule in the vicinity of those active sites). Signal 2 (intermediate **E**) decays to signal 4 (assigned to intermediate **F**) with a significant sKIE, as expected for the conversion of **E** → **F**. Signal 3 also decays to 4, with weak sKIE that suggests proton-independent structural changes. Signal 4 and 1 are similar and represent a slightly different electronic environment of the same species and both decay to signal 5 in parallel. The similarity between signal 5 and that from P450scc with added product (22-hydroxycholesterol product) suggests 5 is the product-bound intermediate; this is further supported by its similarity to the final annealing product. The carbon-centred radical signals resulting from the cryoreduction process ($g \approx 2.1$ to $g \approx 1.95$) have been removed for clarity, and the EPR signal from residual low-spin resting state ferric P450scc was subtracted. Conditions: $T = 25$ K, $\nu = 9.360$ GHz, 13 W microwave power, 10 G modulation amplitude. Adapted with permission from [34]. Copyright © (2012) American Chemical Society

data collected on an oxy-P450scc + cholesterol sample prepared by cryoreduction and subsequent annealing.

Radiolysis introduces a complication; carbon-centred radicals ($g \approx 2$) obscure part of the EPR spectrum and additional unwanted mobile electrons may be able to reduce reactive species in an uncontrolled way. Both these factors appear to be particularly detrimental in the use of cryoreduction techniques to trap and then identify compound I. Additionally, extensive cryoreduction can lead to unwanted double reduction of the heme.

Since compound I has not been directly observed in P450 cryoreduction experiments, its existence and involvement in the reaction mechanism using this technique (prior to Green's work above) was inferred by examining the isotopic composition of the product bound intermediate **I** using deuterated solvents, and substrates, and analysis of ^1H and D (^2H) ENDOR data [11, 31, 34, 35]. The general approach is to compare spectra from samples with either $D_2O/^1H$-substrate or H_2O/D-substrate. Figure 13 shows how the product bound intermediate is expected to be formed for the cases where either **F** or **G** is the catalytically active intermediate.

If the hydroperoxy (**F**) is the reactive intermediate (Fig. 13, Pathway 1) the expected product-bound intermediate **I'** using a ^1H_2O/D-substrate sample will consist of a water molecule (^1H_2O) coordinated to the Fe^{III} heme and a scrambled substrate (^1H/D) hydroxyl proton. The water ^1H nuclei are distinctive in an ENDOR (e.g. Fig. 14) or a HYSCORE spectrum (e.g. Fig. 5 and Fig. 9), and the bound product molecule remote to the Fe^{III} ion will show small ^1H/D hyperfine couplings. If compound I (**G**) is the reactive intermediate (Fig. 13, Pathway 2), the product-bound intermediate **I** is detected in an ENDOR (or HYSCORE) spectrum [11, 31, 32, 34, 35] by a strong deuterium signal intensity and a corresponding loss of the proton signal intensity from the bound hydroxyl. An example of data from an ENDOR study on P450scc is given in Fig. 14 which showed that, as expected, Compound I (intermediate **G**) mediated the hydroxylation.

Measurement of the solvent kinetic isotope effect (sKIE) from samples prepared in both protonated (^1H_2O) or deuterated (D_2O) water has additionally aided in inferring the reactive catalytic species. If **F** or **G** (but not **E**) is the reactive intermediate, there should be a significant sKIE on the rate of product formation [10, 11, 31, 32, 34, 35, 78]. The kinetic data can be obtained by measuring the CW EPR signal intensity at a *g*-value of the desired species as a function of time at a suitably elevated temperature, to allow the reaction to proceed at an observable rate (e.g. 185 K). Note that in P450s and closely related model compounds, there have been no EPR studies that implicate **E** as the active intermediate in monooxygenation reactions, but **E** has been suggested as an active intermediate in P450s able to catalyse lyase activity (CYP17A1) and deformylation reactions (CYP51) [79–82]. If there is a large sKIE in the rate of bound product formation from intermediate **F**, either **F** or **G** must be involved but these two species cannot be distinguished easily through the sKIE alone since both rely upon proton transfer [10, 11, 31, 32, 34, 35, 78].

The following is a collection of selected studies employing cryoreduction to study catalytic reactions carried out by P450 enzymes. We also briefly mention, for comparison, cryoreduction studies of related enzymes.

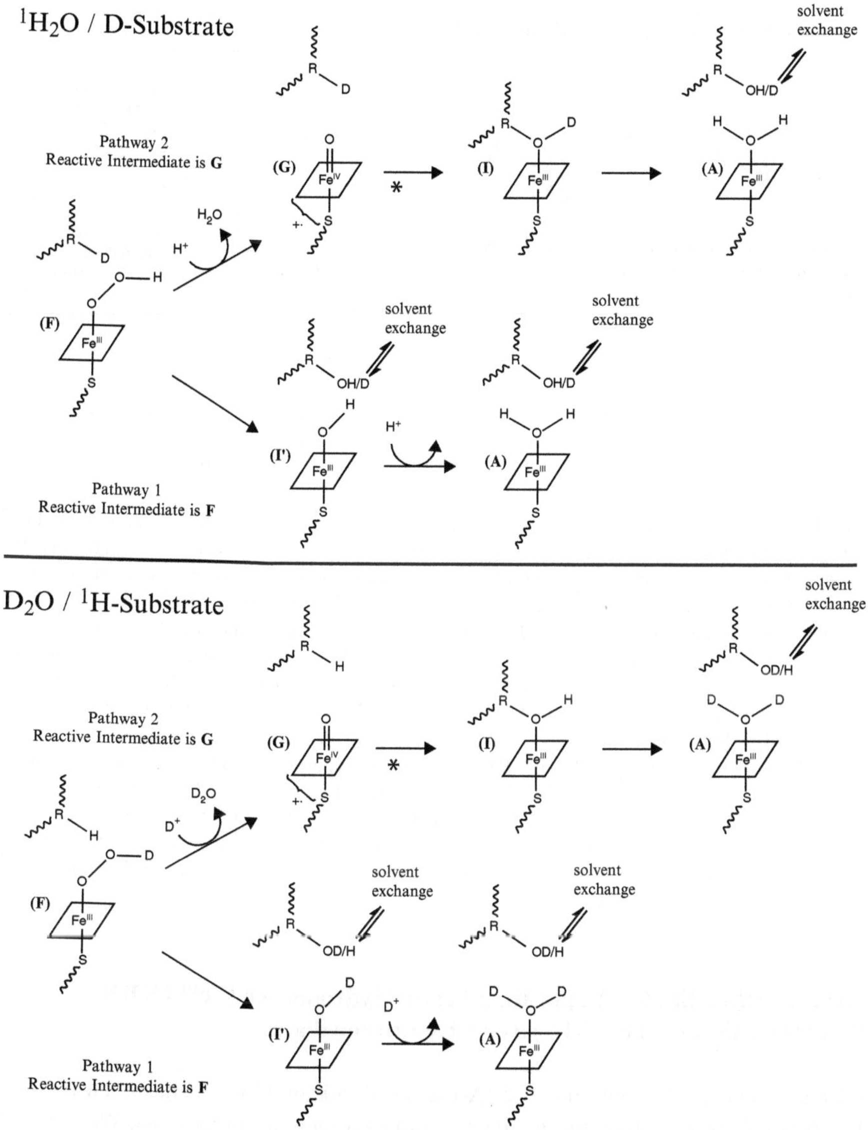

Fig. 13 Two possible reaction pathways for heme monooxygenases that use different reactive intermediates which can be distinguished using two sample preparations [11]; (*top*) deuterated substrate R-D in protonated solvent H_2O, and (bottom) ^1H substrate in deuterated solvent D_2O. ENDOR and/or HYSCORE can be used to measure D or ^1H hyperfine couplings. In Pathway 1, the hydroperoxy (**F**) is the reactive intermediate; there is little sKIE observed in the substrate conversion branch (**F** → **I'**) and the reacting product molecule is not coordinated to the heme iron (intermediate **I'**) and consequently the hydroxyl moiety can exchange its ^1H or D with the solvent. In Pathway 2, compound I (**G**) is the reactive intermediate; there is a large sKIE observed in the substrate conversion step (**G** → **I**) and the reacting product molecule is coordinated to the heme iron and thus the hydroxyl ^1H or D does not exchange with the solvent. The * indicates compound II which is not shown for clarity, see rebound mechanism of Fig. 1

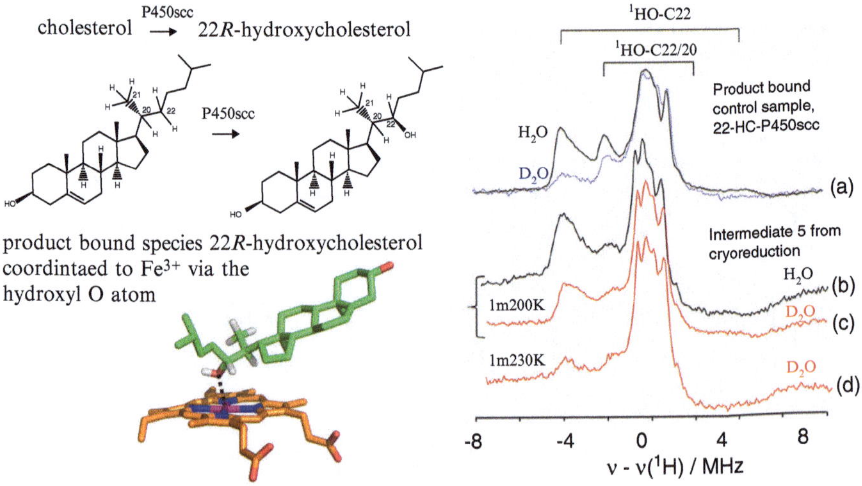

Fig. 14 For the reaction cholesterol → 22*R*–hydroxycholesterol, cryoreduction and Q-band (35 GHz) CW ENDOR at 2 K show that the intermediate **5** formed after cyroreduction (Fig. 12) is the product bound intermediate **I**, and provides evidence that Compound I is the active intermediate. (a) ENDOR data from a sample where the reaction product 22*R*–hydroxycholesterol (22-HC) is added to resting state P450scc (denoted 22-HC-P450scc), allowing its spectral characteristics to be established. Two proton signals are resolved, one from the nearby C22/C20, and one from the hydroxyl proton of the C22 moiety which is exchangeable as determined by the D_2O solvent experiment. (b/c) ENDOR data on the final intermediate **5** (cyroreduction of P450scc + cholesterol then annealing for 1 minute at 200 K). Comparison of these data with those in (a) indicates **5** is the product-bound intermediate, **I** (22-HC- P450scc). When **5** is prepared in D_2O, the proton peak for the hydroxyl proton of coordinated C22-HC is undiminished, as expected if Compound I is the active intermediate. (d) After **5** annealing at 230 K for 1 minute, the HO-C22 proton peak is diminished, indicating that proton exchange can occur at higher temperatures, consistent with the data in (1) and assignment of **5** to bound product **I**. Adapted with permission from [34]. Copyright © (2012) American Chemical Society

Cryoreduction Studies Implicating Ferric-Hydroperoxo (Fe^{III}-OOH) and Ferric Peroxo (Fe^{III}-OO^-) as Active Intermediates

Of the known P450s, only the T252A-mutant P450cam [11, 83] has been reported to employ **F** as the active intermediate of monooxygenation reactions. We note that sulfoxidation and epoxidation reactions carried out by P450s have been studied, but with methods other than EPR, and this work is outside the scope of this review [83–86]. Currently, there is debate over the mechanism of the third reaction of the aromatase cycle, the final lyase step, with EPR evidence inconclusive and computational evidence pointing towards involvement of **E** [23, 87–89].

Mutant P450cam: While wild-type P450cam utilises compound I (**G**) for the hydroxylation of camphor [14], and for the epoxidation of 5-methylenylcamphor (with the product found to be bound to the heme in **I** as determined by EPR) [12], the T252A mutant of P450cam employs species **F**, as shown by observation of distinctive water-coordinated Fe^{III} heme signals [11, 83]. Interestingly, while the D251N mutant of P450cam exhibits a disrupted proton delivery network [9, 10] (and thus is valuable for comparisons between **E** and **F**), a D251N/T252A mutant of P450cam remains able to hydroxylate camphor, despite the disruptions to the formation of **I** caused by T252A [15]. This suggests more than one oxidising state is available to P450cam; **F** could exhibit variable reactivity depending on the mutation T252A or quantities of **G** may still be formed despite the disrupted proton delivery network in the T252A mutant [9–12, 15, 83]. It is also possible that T252A destabilises the iron oxy intermediates.

Aromatase Lyase Step: The aromatase P450 catalytic cycle consists of two sequential hydroxylations of the substrate androstenedione (AD) and a final deformylation (lyase) reaction to give oestrone, i.e., AD → 19-OH-AD → 19-aldo-AD → oestrone. Following the cryoreduction and EPR work of Gantt et al. [23], the first two hydroxylations are suggested to employ **G** as the active intermediate [11, 23], supported by further sKIE studies [90].

The lyase step, however, involves different chemistry in the same active site, leading the authors to consider alternative mechanisms. The cryoreduction study only trapped the ferric peroxo state **E**; the ferric hydroperoxo state was unobserved, unlike for other P450s, perhaps suggesting a hindered proton transfer pathway in the lyase step. Davydov and Hoffman noted that the decrease in intermediate **E** signal was not matched by a synchronous increase in bound product **I** signal, and thus surmised that another intermediate was involved which was not observed in these experiments [11]. Subsequent (non-EPR) sKIE investigation by Khatri et al. [90] indeed showed the lyase step of aromatase catalysis had a similar sKIE to the aromatase hydroxylations, presumably indicating the existence of Compound I in the reactive cycle. With such ambiguous experimental evidence, recent approaches to the problem have come from QM/MM computational studies [87, 88], and Sgrignani et al. [89] have proposed a mechanism where **E** facilitates a nucleophilic attack on the substrate to begin the modification, followed by generation of Compound I that then abstracts a hydrogen from the substrate *via* a water molecule. This proposed mechanism is congruent with the observed sKIE. CYP17 is thought to function via a similar mechanism, see for example [79–82].

Related Non P450 Systems

Nitric oxide synthases are thiolate heme oxygenases that operate in a similar manner to P450s, making them valuable as model systems to further understand P450 catalysis. Two NOSs in particular have been studied with EPR, gsNOS from *Geobacillus stearothermophilus*, and endothelial NOS, eNOS. NOSs are purported to perform two catalytic reactions to create nitric oxide (NO) from arginine the first of which consists of hydroxylation of L-arginine to *N*-hydroxyl-L-arginine (NOHA), and the second of which converts NOHA to citrulline and NO [11, 66, 67]. A short description of selected examples follows.

gsNOS: Unlike in P450s, the **F** intermediate of the second catalytic reaction of gsNOS does not require protonation and so the sKIE can be used to distinguish between **E/F** and **G** [11, 66, 67]. **E** is eliminated as the active intermediate because **F** is observed upon annealing of cryoreduced gsNOS-NOHA. **F** is confirmed as the active intermediate by the lack of a significant sKIE [11].

eNOS: In the second reaction of eNOS catalysis intermediate **E** is trapped following cryoreduction. This leaves open any of **E**, **F** or **G** as potential candidates for the active intermediate. As yet, however, sKIE analysis (to discriminate between **E** and **F**) and EPR of annealing products (to determine if **I** is water-bound or product-bound ferric heme, and thus distinguish between **F** and **G**) have not been carried out [11].

HO: Heme oxygenase (HO) is a histidine ligated enzyme related to P450s that catalyses the degradation of heme to biliverdin, Fe, and CO, in a process that involves HO as both a substrate and a prosthetic group. The heme hydroxylation catalysed by HO is an example where intermediate **F** is the active intermediate, with EPR observation of **F**, a sKIE and mutation studies on cryoreduced HO eliminating **E** as a potential active species [11, 71, 72]. Synchronous conversion of **F** to the α-meso-hydroxyheme during progressive annealing confirms the absence of **G**. Two mutations (D140A or D140F) were shown to disrupt the water network, preventing effective proton transfer and thus preventing normal product formation, resulting in an intermediate analogous to compound II observed in chloroperoxidase [68].

Cryoreduction Studies Implicating Compound I as an Active Intermediate

P450 cryoreduction studies have been successful at trapping intermediates **E** and **F**, and inferring (as described above) that compound I (**G**) is responsible for catalysis. A selection of these studies is listed below.

P450cam: Cryoreduction studies of the P450cam-catalysed conversion of camphor to 5-*exo*-hydroxycamphor [35] have inferred that **G** is the active intermediate. CW EPR spectroscopy data, characterising the kinetics of intermediates during annealing, show a large sKIE when exchanging $^{1}H_2O$ solvent for D_2O, and ^{1}H ENDOR data show the clear absence of protons from water coordinated to the Fe^{III} heme in the bound product intermediate **I**. ENDOR of **I** performed with deuterated substrate discovered deuterium hyperfine couplings consistent with a product-bound heme. These results infer a compound I reactive intermediate, **G**.

P450scc: Davydov et al. showed that compound I (**G**) is the active intermediate of the biologically important conversion of cholesterol to pregenolone by CYP11A1 (P450scc) [31, 34]. The cholesterol to pregenlone conversion is completed in two sequential hydroxylations of cholesterol by P450scc (cholesterol → 22(*R*)-hydroxycholesterol [34] → 20*α*,22(*R*)-dihydroxylcholesterol [31]) and a final cleavage of the C20–C22 bond (producing pregnenolone and isocaproic aldehyde) [31]. Cyroreduction combined with CW EPR was used to identify intermediate species and measure the sKIE, and ^{1}H ENDOR was used to confirm the presence and chemical identity of protons and deuterons near the heme centre at various points during the catalytic cycle. ENDOR signals from ^{14}N ($I = 1$) nuclei were used to probe the conformation of the heme ring to provide information about the binding (or lack of binding) between substrate and heme during catalysis. Altogether, the measurements are consistent with all three reactions being catalysed by **G**.

CYP2B4: In the catalytic cycle of CYP2B4 with substrate butylated hydroxytoluene (BHT) the rapid decay of **I** at the high annealing temperatures needed to progress the reaction makes identification of the reactive intermediate difficult to determine *via* analysis of the product-bound intermediate **I** [78]; in this case, comparison of ^{1}H–substrate/D-solvent and D-substrate/^{1}H–solvent samples cannot be made. However, the authors postulate that the presence of **G** can be inferred by examining the products of cyroreduction and annealing in the absence of substrate. Thus, if **F** reacts directly without involving **G** in the absence of substrate, H_2O_2 will be formed upon annealing [11]. Since H_2O_2 dissociates from the heme rapidly, the resulting system will be a pentacoordinate Fe^{III} heme exhibiting a distinctive high-spin CW EPR spectrum. If the reaction proceeds as **F** → **G** → **I**, H_2O will instead be formed upon subsequent annealing, resulting in a hexacoordinate Fe^{III} heme with bound water, exhibiting a low-spin CW EPR spectrum with distinctive water protons measurable with HYSCORE and/or ENDOR spectroscopy. Thus, the authors suggest that observation of a water-bound, low-spin Fe^{III} spectrum for **I** in substrate-free CYP2B4 experiments implies the presence of **G** in the catalytic cycle [11, 78].

Studies on the F429H proximal loop mutant of CYP2B4 by Davydov and coworkers [32] investigated the effect of hydrogen bonding from the introduced His429 to the thiol of the proximal heme ligand, Cys436. This hydrogen bond significantly perturbed the electronic structure of the Fe^{III} heme (as evidenced by both low- and high-spin CW EPR signals, and ^{14}N heme hyperfine couplings). Throughout the catalytic cycle (progressed by annealing) of the F429H-CYP2B4 mutant, the decay rate reaction kinetics of the ferric hydroperoxy intermediate **F** was significantly slowed which reduces the efficiency of substrate hydroxylation in the mutant enzyme by the active intermediate, compound I (**G**). This data underscores the influence of ligand effects on the reactivity of P450 intermediates. In F429H-CYP2B4 the hydrogen bond decreases the extent of S → Fe electron donation and weakens the Fe-S bond. The authors postulate this change in electronic structure diminishes the reactivity of compound I by changing the pK_a of the $Fe^{IV} = O$ moiety (see Eq. (11)) [32].

It is also noteworthy that cyroreduction without annealing generally traps the parent molecular configuration. For example, for the parent Fe(II) WT-CYP2B4 enzyme (prepared as intermediate **D**) the electron from cyroreduction becomes localised on the porphyrin generating an $S = 3/2$ spin-state with the anion radical ($S = 1/2$) exchanged coupled antiferromagnetically to the Fe^{II} ion ($S = 2$). Further details are given in the original paper [32].

Related Non P450 Systems

gsNOS: In the first stage of gsNOS catalysis (see above) where the transition from **G** → **I** is sufficiently fast that only **I** is observed, EPR and ^{15}N ENDOR data show the product bound intermediate has NOHA in complex with Fe^{III} heme (rather than with water), providing indirect evidence for Compound I (**G**) as the reactive intermediate [66, 67].

eNOS: The first catalytic stage of eNOS (see above) provides another example where **I** can be used to distinguish between **F** and **G** as the reactive species; unlike gsNOS, eNOS is found only as **E** following cryoreduction, and **F** and **G** are not observed during annealing. However, the EPR and ENDOR spectra for the product bound intermediate **I** of eNOS are similar to those for **I** of gsNOS which suggests the eNOS catalytic reaction proceeds through the active intermediate **G**. The authors hypothesised the difference in cryoreduction products between eNOS and gsNOS to be due to differences in their hydrogen-bonding networks, with eNOS lacking a nearby ordered water molecule until annealing allows water to approach the ferric peroxo-heme (**E**), and produced intermediate **F** [66, 67].

P450 Conformations and Electron Transfer Complexes by Dipolar EPR (DEER)

The crystal structures of P450s have shown that some enzymes have access to a range of conformers—a closed active site conformer, and several open active site conformers. How the structure of these conformations control substrate binding and specificity is of great functional interest and thus techniques which can probe conformational heterogeneity in solution are valuable. The DEER technique is ideally suited for this task—the protein can be labelled with two paramagnetic spin probes and the distance distribution between the labels measured in solution (frozen solution) under a variety of conditions by DEER spectroscopy. In the paper by Stoll, [41] this technique was applied to P450cam which has crystal structures available that reveal one closed and several open conformations. To monitor protein conformation with and without substrate a set of carefully chosen spin label pairs were used to monitor regions of the protein where structural changes are predicted in going from an open to the closed conformation. An example of the DEER experimental data is shown in Fig. 15 for the P450cam double mutant, S48C/S190C, in which the introduced cysteines were labelled with MTSSL. Figure 15 shows two time-domain DEER traces collected for substrate free and camphor-bound P450cam, and the corresponding distance distributions computed by Tikhonov regularisation. The data show a clear reduction in the label distance of *ca.* 8 Å when the substrate camphor binds. From the set of experimental DEER distance distributions the authors were able to conclude, in a physiologically relevant solution state, that upon binding of the substrate, P450cam goes from an open to a closed active site conformation. Using the same DEER methodology Myers et al. [42] were able to characterise P450cam structural changes when complexed with putidaredoxin and observe a dependence upon the P450 oxidation state [41, 42].

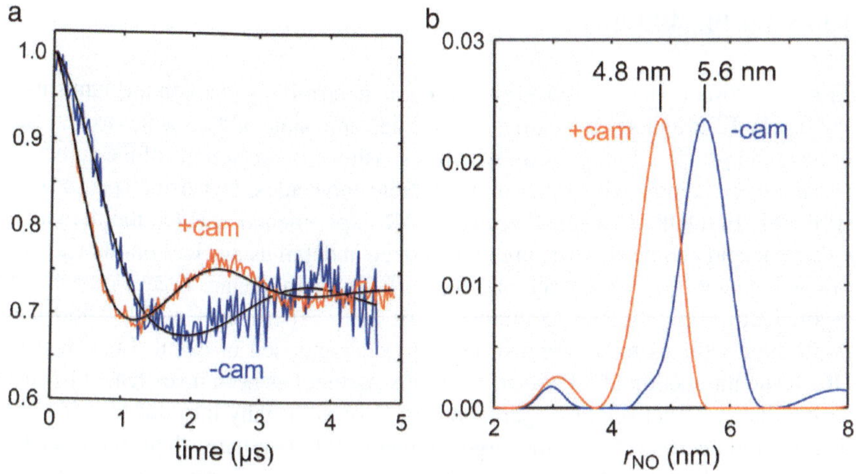

Fig. 15 DEER traces showing that a conformation changes occurs when P450cam binds its substrate camphor. *Left*: DEER traces with (*red*) and without camphor (*blue*) and the fits (*black*) that give the DEER distributions (*right*) which are computed by Tikhonov regularisation. Adapted with permission from reference [41]. Copyright © (2012) American Chemical Society

DEER can also be used to obtain structural information on protein complexes of the electron transport chain. P450 catalysis is dependent upon the active site receiving two electrons sequentially from NAD(P)H usually via one or two electron transfer proteins. In the paper by Lovett et al. [45] preliminary data for the electron transfer complex formed between palustrisredoxin reductase (PuR) and palustrisredoxin (Pux-B) from *Rhodopseudomonas palustris* were presented. The DEER experiments utilised the reduced $[2Fe–2S]^{+\bullet}$ cluster in Pux-B (spin $S = 1/2$) with an MTSSL spin label on PuR. The DEER data show that distance and orientation information between the two paramagnetic probes can be obtained and thus a set of such measurements can be used to define the structure of a complex. Although many P450s and a handful of redox partners have available crystal structures, until recently there was little information available on the structure of redox couples. Poulos provided the first redox couple crystal structures with covalently linked complexes of Putidaredoxin Reductase:Putidaredoxin (PdR-Pdx) [91] and P450cam: putidaredoxin (P450cam-Pdx) [92]. This was followed immediately by X-ray and paramagnetic NMR structures of a P450 mutant but, significantly, without possible biases from the covalent link between the proteins [46, 93]. The combination of techniques described above offers the possibility of probing structural features in a range of P450s and identifying critical structural elements which control partner recognition and efficient electron transfer [45, 46, 91].

Concluding Remarks

EPR spectroscopy has played a critical role in determining reaction mechanisms of P450. CW EPR enables the oxidation state and spin state of the species to be determined, and provides a fingerprint which is sensitive to interactions of the active site paramagnetic centre with water and substrate molecules. Hyperfine spectroscopy (ESEEM, ENDOR, ELDOR-detected NMR experiments) enables the active site electronic and geometric structure to be probed in detail using electron-nuclear couplings between e.g. a substrate and the Fe^{III} ion of the heme. Freeze-quench and cryoreduction experiments combined with EPR experiments enable short-lived P450 intermediates to be trapped and thus characterised in detail. Finally, using DEER the interaction of a P450 with a redox partner has been investigated by measuring distances between two paramagnetic probes, usually introduced using site-directed mutagenesis. Given the large number of P450 systems, their importance in biology, and the desire to reengineer them to be utilised in a range of reactions for chemical synthesis of target molecules, the methods discussed in this review will continue to play an important role in P450 research into the future.

References

1. Poulos TL (2014) Heme enzyme structure and function. Chem Rev 114(7):3919–3962
2. Girvan HM, Munro AW (2016) Applications of microbial cytochrome P450 enzymes in biotechnology and synthetic biology. Curr Opin Chem Biol 31:136–145
3. Bernad S et al (2012) A novel cryo-reduction method to investigate the molecular mechanism of nitric oxide synthases. J Phys Chem B 116(19):5595–5603
4. Munro AW et al (2013) What makes a P450 tick? Trends Biochem Sci 38(3):140–150
5. de Montellano PRO (2010) Hydrocarbon hydroxylation by cytochrome P450 enzymes. Chem Rev 110(2):932–948
6. Stok JE et al (2013) Cytochrome $P450_{cin}$ (CYP176A1) D241N: Investigating the role of the conserved acid in the active site of cytochrome P450s. Biochim Biophys Acta Prot Proteom 1834(3):688–696
7. Johnston WA et al (2011) Cytochrome P450 is present in both ferrous and ferric forms in the resting state within intact Escherichia coli and hepatocytes. J Biol Chem 286(47):40750–40759
8. Conner KP et al (2012) 1,2,3-triazole-heme interactions in cytochrome P450: functionally competent triazole-water-heme complexes. Biochemistry 51(32):6441–6457
9. Davydov R et al (1999) EPR and ENDOR of catalytic intermediates in cryoreduced native and mutant oxy-cytochromes P450cam: mutation-induced changes in the proton delivery system. J Am Chem Soc 121(45):10654–10655
10. Davydov R et al (2001) Hydroxylation of camphor by-reduced oxy-cytochrome P450cam: mechanistic implications of EPR and ENDOR studies of catalytic intermediates in native and mutant enzymes. J Am Chem Soc 123(7):1403–1415
11. Davydov R, Hoffman BM (2011) Active intermediates in heme monooxygenase reactions as revealed by cryoreduction/annealing, EPR/ENDOR studies. Arch Biochem Biophys 507(1):36–43
12. Kim SH et al (2005) Cryoreduction EPR and ^{13}C, ^{19}F ENDOR study of substrate-bound substates and solvent kinetic isotope effects in the catalytic cycle of cytochrome P450cam and its T252A mutant. Dalton Trans 21:3464–3469

13. Rittle J, Green MT (2010) Cytochrome P450 compound I: capture, characterization, and C-H bond activation kinetics. Science 330(6006):933–937
14. Yosca TH, Green MT (2016) Preparation of compound I in P450cam: the prototypical P450. Israel J Chem 56(9–10):834–840
15. Denisov IG et al (2005) Structure and chemistry of cytochrome P450. Chem Rev 105(6):2253–2278
16. Hofrichter M et al (2015) Fungal unspecific peroxygenases: heme-thiolate proteins that combine peroxidase and cytochrome P450 properties. In: Hrycay EG, Bandiera SM (eds) Monooxygenase, peroxidase and peroxygenase properties and mechanisms of cytochrome P450. Springer International Publishing, Cham, pp 341–368
17. Hrycay EG, Bandiera SM (2015) Monooxygenase, peroxidase and peroxygenase properties and reaction mechanisms of cytochrome P450 enzymes. In: Hrycay EG, Bandiera SM (eds) Monooxygenase, peroxidase and peroxygenase properties and mechanisms of cytochrome P450. Springer International Publishing, Cham, pp 1–61
18. Shoji O, Watanabe Y (2014) Peroxygenase reactions catalyzed by cytochromes P450. J Biol Inorg Chem 19(4):529–539
19. Krest CM et al (2013) Reactive intermediates in cytochrome P450 catalysis. J Biol Chem 288(24):17074–17081
20. Hannemann F et al (2007) Cytochrome P450 systems—biological variations of electron transport chains. Biochim Biophys Acta 1770(3):330–344
21. Tsai R et al (1970) Spin-state changes in cytochrome P-450$_{cam}$ on binding of specific substrates. Proc Natl Acad Sci U S A 66(4):1157–1163
22. Kagawa N et al (2004) Characterization of stable human aromatase expressed in E-coli. Steroids 69(4):235–243
23. Gantt SL et al (2009) The critical iron-oxygen intermediate in human aromatase. Biochem Biophys Res Commun 387(1):169–173
24. Maurelli S et al (2011) Direct spectroscopic evidence for binding of anastrozole to the iron heme of human aromatase. Peering into the mechanism of aromatase inhibition. Chem Commun 47(38):10737–10739
25. Sligar SG (1976) Coupling of spin, substrate, and redox equilibria in cytochrome P450. Biochemistry 15(24):5399–5406
26. Bohan TL (1977) Analysis of low-spin ESR spectra of ferric heme proteins: a reexamination. J Magn Reson 26(1):109–118
27. Taylor CPS (1977) EPR of low-spin heme complexes–relation of t_{2g} hole model to directional properties of g-tensor, and a new method for calculating ligand-field parameters. Biochim Biophys Acta 491(1):137–149
28. Conner KP et al (2014) Strength of axial water ligation in substrate-free cytochrome P450s is isoform dependent. Biochemistry 53(9):1428–1434
29. Takeuchi K et al (2001) Adrenodoxin-cytochrome P450scc interaction as revealed by EPR spectroscopy: comparison with the putidaredoxin-cytochrome P450cam system. J Biochem 130(6):789–797
30. Schweiger A, Jeschke G (2001) Principles of pulse electron paramagnetic resonance. Oxford University Press, Inc., New York, NY
31. Davydov R et al (2015) Evidence that compound i is the active species in both the hydroxylase and lyase steps by which P450scc converts cholesterol to pregnenolone: EPR/ENDOR/Cryoreduction/Annealing studies. Biochemistry 54:7089–7097
32. Davydov R et al (2016) Role of the proximal cysteine hydrogen bonding interaction in cytochrome P450 2B4 studied by cryoreduction, electron paramagnetic resonance, and electron-nuclear double resonance spectroscopy. Biochemistry 55(6):869–883
33. Davydov R et al (2010) Probing the oxyferrous and catalytically active ferryl states of Amphitrite ornata dehaloperoxidase by cryoreduction and EPR/ENDOR spectroscopy. Detection of compound I. J Am Chem Soc 132(42):14995–15004

34. Davydov R et al (2012) Compound I is the reactive intermediate in the first monooxygenation step during conversion of cholesterol to pregnenolone by cytochrome P450scc: EPR/ENDOR/Cryoreduction/Annealing studies. J Am Chem Soc 134(41):17149–17156
35. Davydov R et al (2013) The use of deuterated camphor as a substrate in ^{1}H ENDOR studies of hydroxylation by cryoreduced oxy P450cam provides new evidence of the involvement of compound I. Biochemistry 52(4):667–671
36. McQuarters AB et al (2014) 1958–2014: after 56 years of research, cytochrome P450 reactivity is finally explained. Angew Chem Int Ed 53(19):4750–4752
37. Groves JT (2014) Enzymatic C-H bond activation: using push to get pull. Nat Chem 6(2):89–91
38. Krest CM et al (2015) Significantly shorter Fe–S bond in cytochrome P450-I is consistent with greater reactivity relative to chloroperoxidase. Nat Chem 7(9):696–702
39. Timmel CR, Harmer JR (2013) Spin-labels and intrinsic paramagnetic centres in the Biosciences: structural information from distance measurements. In: Mingos DMP (ed) Structure & Bond, vol 152. Springer-Verlag Berlin, Heidelberg, p 322
40. Bowen AM et al (2013) Orientation-selective DEER using rigid spin labels, cofactors, metals, and clusters. In: Structure and bonding: structural information from spin-labels and intrinsic paramagnetic centres in the biosciences, vol 152, pp 283–327
41. Stoll S et al (2012) Double electron-electron resonance shows cytochrome P450cam undergoes a conformational change in solution upon binding substrate. Proc Natl Acad Sci U S A 109(32):12888–12893
42. Myers WK et al (2013) The conformation of P450cam in complex with putidaredoxin is dependent on oxidation state. J Am Chem Soc 135(32):11732–11735
43. Ezhevskaya M et al (2013) Distance determination between low-spin ferric haem and nitroxide spin label using DEER: the neuroglobin case. Mol Phys 111:2855–2864
44. Abdullin D et al (2015) Comparison of PELDOR and RIDME for distance measurements between nitroxides and low-spin Fe(III) ions. J Phys Chem B 119(43):13534–13542
45. Lovett JE et al (2009) Structural information from orientationally selective DEER spectroscopy. Phys Chem Chem Phys 11(31):6840–6848
46. Hiruma Y et al (2013) The structure of the cytochrome P450cam-putidaredoxin complex determined by paramagnetic NMR spectroscopy and crystallography. J Mol Biol 425(22):4353–4365
47. Peisach J, Blumberg WE (1970) Electron paramagnetic resonance study of the high- and low-spin forms of cytochrome P-450 in liver and in liver microsomes from a methylcholanthrene-treated rabbit. Proc Natl Acad Sci 67(1):172–179
48. Salerno JC et al (1995) Characterization by electron paramagnetic resonance of the interactions of L-arginine and L-thiocitrulline with the heme cofactor region of nitric oxide synthase. J Biol Chem 270(46):27423–27428
49. Salerno JC, McMillan K, Masters BSS (1996) Binding of intermediate, product, and substrate analogs to neuronal nitric oxide synthase: ferriheme is sensitive to ligand-specific effects in the L-arginine binding site. Biochemistry 35(36):11839–11845
50. Solomon EI et al (1997) New insights from spectroscopy into the structure/function relationships of lipoxygenases. Chem Biol 4(11):795–808
51. Telser J, Krzystek J, Ozarowski A (2014) High-frequency and high-field electron paramagnetic resonance (HFEPR): a new spectroscopic tool for bioinorganic chemistry. J Biol Inorg Chem 19(3):297–318
52. Abragam A, Bleaney B (1970) Electron paramagnetic resonance of transition ions. Oxford University Press, Oxford
53. Epel B et al (2001) The effect of spin relaxation on ENDOR spectra recorded at high magnetic fields and low temperatures. J Magn Reson 148(2):388–397
54. Morton JR, Preston KF (1978) Atomic parameters for paramagnetic resonance data. J Magn Reson 30(3):577–582
55. Goldfarb D et al (1996) Study of water binding to low-spin Fe(III) in cytochrome P450 by pulsed ENDOR and four-pulse ESEEM spectroscopies. J Am Chem Soc 118(11):2686–2693

56. Ye S, Neese F (2009) Quantum chemical studies of C-H activation reactions by high-valent nonheme iron centers. Curr Opin Chem Biol 13(1):89–98
57. Hirao H, Thellamurege N, Zhang X (2014) Applications of density functional theory to iron-containing molecules of bioinorganic interest. Front Chem 2:26
58. Shaik S et al (2010) P450 enzymes: their structure, reactivity, and selectivity—modeled by QM/MM calculations. Chem Rev 110(2):949–1017
59. Luthra A, Denisov IG, Sligar SG (2011) Spectroscopic features of cytochrome P450 reaction intermediates. Arch Biochem Biophys 507(1):26–35
60. Cruce AA, Lockart M, Bowman MK (2015) Chapter twelve–pulsed EPR in the study of drug binding in cytochrome P450 and NOS. In: Peter ZQ, Kurt W (eds) Methods in enzymology. Academic Press, Cambridge, MA, pp 311–340
61. García-Rubio I et al (2003) HYSCORE spectroscopy in the cytochrome b559 of the photosystem II reaction center. J Am Chem Soc 125(51):15846–15854
62. Thomann H et al (1995) Evidence for water binding to the Fe centre in cytochrome P450cam obtained by ^{17}O electron-spin echo envelope modulation spectroscopy. J Am Chem Soc 117(31):8243–8251
63. Roberts AG et al (2010) Intramolecular heme ligation of the cytochrome P450 2C9 R108H mutant demonstrates pronounced conformational flexibility of the B-C loop region: implications for substrate binding. Biochemistry 49(40):8700–8708
64. Van Heuvelen KM et al (2012) One-electron oxidation of an oxoiron(IV) complex to form an $[O{=}Fe^{V}{=}NR]^{+}$ center. Proc Natl Acad Sci U S A 109(30):11933–11938
65. Yosca TH et al (2013) Iron(IV)hydroxide pK_a and the role of thiolate ligation in C–H bond activation by cytochrome P450. Science 342(6160):825–829
66. Davydov R et al (2009) EPR and ENDOR characterization of the reactive intermediates in the generation of NO by cryoreduced oxy-nitric oxide synthase from geobacillus stearothermophilus. J Am Chem Soc 131(40):14493–14507
67. Davydov R et al (2002) EPR and ENDOR characterization of intermediates in the cryoreduced oxy-nitric oxide synthase heme domain with bound l-arginine or NG-hydroxyarginine. Biochemistry 41(33):10375–10381
68. Rutter R et al (1984) Chloroperoxidase compound I: electron paramagnetic resonance and Moessbauer studies. Biochemistry 23(26):6809–6816
69. Wang X et al (2013) Driving force for oxygen-atom transfer by heme-thiolate enzymes. Angew Chem Int Ed 52(35):9238–9241
70. Wang X et al (2015) Heme-thiolate ferryl of aromatic peroxygenase is basic and reactive. Proc Natl Acad Sci 112(12):3686–3691
71. Davydov R et al (2002) Catalytic mechanism of heme oxygenase through EPR and ENDOR of cryoreduced oxy-heme oxygenase and its Asp 140 mutants. J Am Chem Soc 124(8):1798–1808
72. Davydov R et al (2003) Kinetic isotope effects on the rate-limiting step of heme oxygenase catalysis indicate concerted proton transfer/heme hydroxylation. J Am Chem Soc 125(52):16208–16209
73. Zilly FE et al (2011) Tuning a P450 enzyme for methane oxidation. Angew Chem Int Ed 50(12):2720–2724
74. Green MT (2009) C–H bond activation in heme proteins: the role of thiolate ligation in cytochrome P450. Curr Opin Chem Biol 13(1):84–88
75. Green MT, Dawson JH, Gray HB (2004) Oxoiron(IV) in chloroperoxidase compound II is basic: Implications for P450 chemistry. Science 304(5677):1653–1656
76. Behan RK, Green MT (2006) On the status of ferryl protonation. J Inorg Biochem 100(4):448–459
77. Warren JJ, Tronic TA, Mayer JM (2010) Thermochemistry of proton-coupled electron transfer reagents and its implications. Chem Rev 110(12):6961–7001
78. Davydov R et al (2008) Characterization of the microsomal cytochrome P450 2B4 O_2 activation intermediates by cryoreduction and electron paramagnetic resonance. Biochemistry 47(36):9661–9666

79. Khatri Y et al (2014) Active site proton delivery and the lyase activity of human CYP17A1. Biochem Biophys Res Commun 443(1):179–184
80. Mak PJ et al (2015) Unveiling the crucial intermediates in androgen production. Proc Natl Acad Sci 112(52):15856–15861
81. Mak PJ et al (2014) Resonance raman spectroscopy reveals that substrate structure selectively impacts the heme-bound diatomic ligands of CYP17. Biochemistry 53(1):90–100
82. Sen K, Hackett JC (2010) Peroxo−iron mediated deformylation in sterol 14α-demethylase catalysis. J Am Chem Soc 132(30):10293–10305
83. Jin S et al (2003) Epoxidation of olefins by hydroperoxo−ferric cytochrome P450. J Am Chem Soc 125(12):3406–3407
84. Cryle MJ, De Voss JJ (2006) Is the ferric hydroperoxy species responsible for sulfur oxidation in cytochrome P450s? Angew Chem Int Edn 45(48):8221–8223
85. Volz TJ, Rock DA, Jones JP (2002) Evidence for two different active oxygen species in cytochrome p450 BM3 mediated sulfoxidation and N-dealkylation reactions. J Am Chem Soc 124(33):9724–9725
86. Ortiz de Montellano PR (2010) Hydrocarbon hydroxylation by cytochrome P450 enzymes. Chem Rev 110(2):932–948
87. Di Nardo G, Gilardi G (2013) Human aromatase: perspectives in biochemistry and biotechnology. Biotechnol Appl Biochem 60(1):92–101
88. Akhtar M, Wright JN, Lee-Robichaud P (2011) A review of mechanistic studies on aromatase (CYP19) and 17α-hydroxylase-17,20-lyase (CYP17). J Steroid Biochem Mol Biol 125(1–2):2–12
89. Sgrignani J, Iannuzzi M, Magistrato A (2015) Role of water in the puzzling mechanism of the final aromatization step promoted by the human aromatase enzyme. Insights from QM/MM MD simulations. J Chem Inf Model 55(10):2218–2226
90. Khatri Y et al (2014) Kinetic solvent isotope effect in steady-state turnover by CYP19A1 suggests involvement of Compound 1 for both hydroxylation and aromatization steps. FEBS Lett 588(17):3117–3122
91. Sevrioukova IF, Poulos TL, Churbanova IY (2010) Crystal structure of the putidaredoxin reductase·putidaredoxin electron transfer complex. J Biol Chem 285(18):13616–13620
92. Tripathi S, Li H, Poulos TL (2013) Structural basis for effector control and redox partner recognition in cytochrome P450. Science 340(6137):1227–1230
93. Skinner SP et al (2015) Delicate conformational balance of the redox enzyme cytochrome P450cam. Proc Natl Acad Sci 112(29):9022–9027
94. Lipscomb JD (1980) Electron paramagnetic resonance detectable states of cytochrome P-450cam. Biochemistry 19(15):3590–3599
95. Groves JT, McClusky GA (1976) Aliphatic hydroxylation via oxygen rebound. Oxygen transfer catalyzed by iron. J Am Chem Soc 98(3):859–861
96. Cryle MJ, Ortiz de Montellano PR, De Voss JJ (2005) Cyclopropyl containing fatty acids as mechanistic probes for cytochromes P450. J Org Chem 70(7):2455–2469

Characterization of Radical S-adenosylmethionine Enzymes and Intermediates in their Reactions by Continuous Wave and Pulse Electron Paramagnetic Resonance Spectroscopies

Alexey Silakov, Nicholas D. Lanz, and Squire J. Booker

Abstract Radical S-adenosylmethionine (SAM) enzymes comprise an important and a versatile superfamily of enzymes. For more than a decade, a significant effort has been directed towards understanding these enzymes. Electron paramagnetic resonance spectroscopy has played a crucial role in such studies, helping to decipher intricate details about the identity of the active metallocofactors, their relations to the substrate(s) utilized and an understanding of the mechanisms of the enzymatic reactions. In this chapter we overview research milestones in the field of radical SAM enzymes achieved with the aid of EPR spectroscopy.

Keywords ENDOR • HYSCORE • ESEEM • CW-EPR • Cfr • RlmN • Lipoyl synthase • Lysine 2,3-aminomutase • HydG • Pyruvate formate-lyase • Glycyl radical • S-adenosylmethionine • Nucleic acid radical • [4Fe–4S] cluster • Biotin synthase • MoaA • Rapid-freeze quench EPR • Methylase

Introduction

In the past decade, a multitude of studies have been performed to interrogate the structures and mechanisms of the radical SAM (RS) superfamily of enzymes, and electron paramagnetic resonance (EPR) spectroscopic techniques have been vital in these endeavors. Members of the RS superfamily of enzymes catalyze an incredibly diverse set of chemical reactions. To date there are almost 114,000 individual sequences that belong to the family, and the products of these genes catalyze forty-eight distinct and known reactions [1, 2].

A. Silakov (✉) • N.D. Lanz • S.J. Booker
Departments of Chemistry and Biochemistry and Molecular Biology, The Howard Hughes Medical Institute, The Pennsylvania State University, University Park, PA 16802, USA
e-mail: alexey.silakov@gmail.com

G. Hanson, L. Berliner (eds.), *Future Directions in Metalloprotein and Metalloenzyme Research*, Biological Magnetic Resonance 33,
DOI 10.1007/978-3-319-59100-1_6

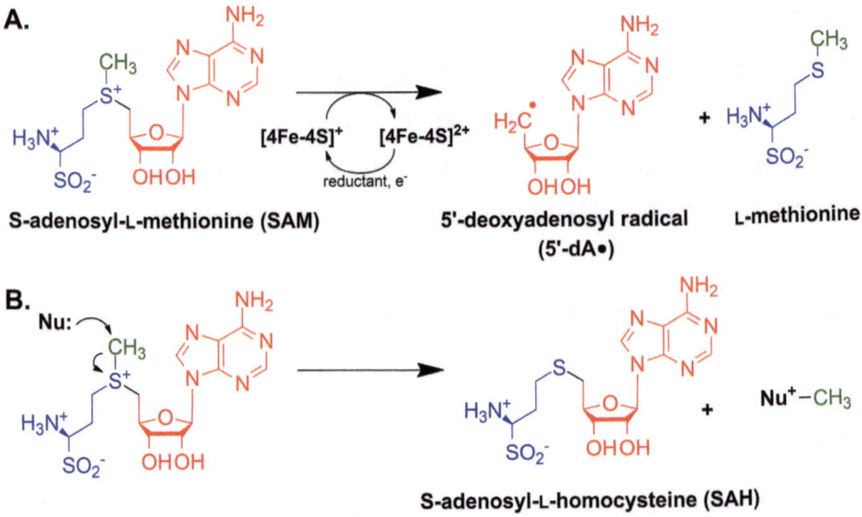

Scheme 1 Reactions of S-adenosyl-L-methionine in radical SAM enzymes. (**a**) Reductive cleavage of SAM into L-methionine and 5'-deoxyadenosyl radical. (**b**) Methylgroup transfer in an S_N2 reaction from SAM to a nucleophile (Nu) with a release of S-adenosyl-L-homocysteine (SAH)

Most RS enzymes are easily identified by a CX_3CX_2C motif found in their primary structures [3]. The three cysteines in this motif coordinate three of the four iron ions of a [4Fe–4S] cluster. The fourth iron ion is not ligated by a protein residue, which allows *S*-adenosylmethionine (SAM) to bind to it in a bidentate fashion via its α-carboxylate and α-amino groups. Binding in this manner positions the sulfur atom of SAM to accept an electron from the reduced $[4Fe–4S]^+$ cluster, resulting in homolytic cleavage of the C5′–S bond of SAM [4–8]. Homolytic cleavage of SAM results in the formation of methionine and a 5′-deoxyadensoyl radical (5′-dA•), which is a potent oxidizing agent. The 5′-dA• is then used to abstract a specific hydrogen atom from a substrate molecule [9] (Scheme 1).

The 5′-dA• is a common tool used by Nature to initiate radical chemistry. This same oxidizing agent is generated by the adenosylcobalamin (vitamin B_{12}) cofactor, which undergoes a reversible homolytic cleavage of the C5′–Co bond to form the 5′-dA• [10]. SAM can also be reversibly cleaved to generate a 5′-dA•, allowing it to be used as a cofactor; however, it is most commonly consumed as a substrate in the reactions in which it participates. Upon abstraction of a hydrogen atom by the 5′-dA•, 5′-deoxyadenosine (5′-dA) and a substrate radical are formed. These substrates can be large biological polymers like proteins, RNA and DNA, or small molecules. In some cases the initial substrate radical formed is the final product of the RS reaction—as in the case of proteins that contain glycyl radical cofactors—but in most cases further bond-breaking and bond-making takes place. The complexity of the reactions catalyzed by RS enzymes varies dramatically from a single hydrogen atom abstraction, as in the case of pyruvate formate–lyase activating enzyme) (PFL-AE), to highly complex rearrangements, as observed in MoaA [9].

RS enzymes are uniquely suited to study by EPR spectroscopy because of their redox active [4Fe–4S] clusters and the presence of radical intermediates in their reactions. Additionally, a number of RS enzymes contain multiple Fe/S clusters that can participate in redox chemistry, bind intermediates, or interact with other EPR active species [11, 12]. Because of these features, EPR spectroscopy can be used to probe the active sites and key intermediates of RS enzymes, which allows for detailed mechanistic analyses that are simply not possible for enzymes that do not contain paramagnetic intermediates.

EPR is an invaluable tool for resolving structural and electronic properties of paramagnetic centers in general and catalytic intermediates in particular. In some cases the resolution of EPR spectra can be increased by a procedure called resolution enhancement (RE), which takes advantage of Fourier deconvolution of Lorentzian or Gaussian lineshapes [13, 14]. First, an inverse Fourier transformation is applied to the original spectrum and followed by a division of the time-domain trace with an exponential function (deconvolution of the Lorentzian lineshape). An apodization window function is then applied and the RE spectrum is obtained by subsequent forward Fourier transformation. The original article by Kauppinen *et al.* mentions $[1-|x|/L]^2$ as an apodization function; however, other common functions, e.g. Hamming, Gaussian etc., can provide similar results. Two input parameters are required for this procedure: the width of the unitary lineshape and the apodization parameter (L). It is important to note that because neither the function nor the linewidth of the unitary line is generally known, the best values must be obtained by trial and error. This necessity makes the outcome dependent on the judgement of the user and thus may cause overinterpretation of the experimental results. A comprehensive overview of this methodology has been published by Reed and Poyner [13].

Advanced EPR spectroscopic techniques provide a much more convenient means to enhance spectroscopic resolution substantially and allow for a deeper view into the fine details of weaker magnetic interactions. Continuous wave Electron Nuclear Double Resonance (CW ENDOR) spectroscopy has been extensively used to obtain extraordinary details on paramagnetic species in RS enzymes and their reactions. The technique takes advantage of the fact that the amplitude of the EPR spectrum will vary if a radio frequency (RF) excitation is resonant with an NMR transition associated with an EPR transition. By scanning radio frequencies, nuclear transitions associated with magnetic nuclei in the vicinity of paramagnetic species can be resolved. The main disadvantage of this technique is that the RF perturbation occurs at the same time as the detection of the EPR signal, which makes this technique very sensitive to heating artifacts. As a result, the experimental setup needs to be considered very carefully. Pulse EPR provides an alternative way of performing ENDOR measurements. In this case, a RF is applied as a pulsed perturbation well before the detection. The two most common pulse methods are Mims and Davies ENDOR. The main disadvantage of pulse ENDOR methods as compared to CW ENDOR is the presence of blind spot patterns that alter the "true" lineshape of the ENDOR signal and, especially for weakly coupled nuclei, may strongly suppress the ENDOR signal. Therefore, application of this technique to weak, purely anisotropic hyperfine (HF) interactions is limited.

Another frequently used double-resonance pulse EPR technique to resolve hyperfine interaction is called ELDOR detected NMR; however, given that, to our knowledge, no application of such methodology to the field of RS enzymes has been reported, this technique will not be covered here.

A second set of pulse EPR techniques, electron spin echo envelope modulation (ESEEM), provides another way of resolving HF interactions without using a second distinct excitation frequency. In this class of techniques, the paramagnetic NMR frequencies are encoded in the dependence of the electron spin echo on a delay between microwave pulses. The limitation of the technique is that the amplitude of the effect depends on the degree of anisotropy of the HF interaction. The technique is most suitable to study weak, anisotropic, hyperfine interactions, which makes it complementary to ENDOR methods. Hyperfine sublevel correlation spectroscopy (HYSCORE) is a variation of ESEEM that allows for correlation of nuclear frequencies on a two-dimensional spectrum. Most importantly, HYSCORE can separate nuclear frequencies for weak and strong HF coupling regimes in different parts of the spectrum, such as when a hyperfine coupling constant (A) is smaller or larger than twice the Larmor frequency ($2\nu_L$).

Detailed explanations of the techniques described within have been provided by others [15, 16]. In what follows, we present several milestones in studying this versatile superfamily of enzymes that were achieved by EPR spectroscopy. We use the case studies of various radical SAM enzymes to highlight how the tools of EPR spectroscopy can be used to (1) identify the forms of Fe/S clusters present and monitor the binding of substrates to the clusters, (2) identify substrate and intermediate radicals that result from catalysis, and (3) use radical intermediates to probe the active sites to determine fine structure and electronic properties.

EPR Study of Metalloclusters in RS Enzymes

Radical SAM-Binding Cluster

The core feature of RS enzymes is the presence of a [4Fe–4S] cluster ($[4Fe–4S]_{RS}$) that is coordinated by three cysteine ligands that are typically arranged in a signature CX_3CX_2C motif. The fourth iron ion is not coordinated by an amino acid ligand and is labile. Most notably, this cluster is sensitive to oxidative degradation, leading to [3Fe–4S] or even [2Fe–2S] constructs. The degraded [3Fe–4S] cluster form is also commonly found in as-isolated RS enzymes, especially if the enzymes are purified aerobically (see Fig. 1, A) [9]. CW EPR spectroscopy has proven to be useful in identifying the forms of Fe/S clusters in RS enzymes and in detecting the binding of SAM to the unique iron site.

The first direct evidence for the unique iron site in the $[4Fe–4S]_{RS}$ cluster was obtained in studies of Lysine 2,3-aminomutase (LAM), wherein its $[4Fe–4S]_{RS}$ cluster could be oxidized to a [3Fe–4S] cluster and then converted back to the $[4Fe–4S]_{RS}$

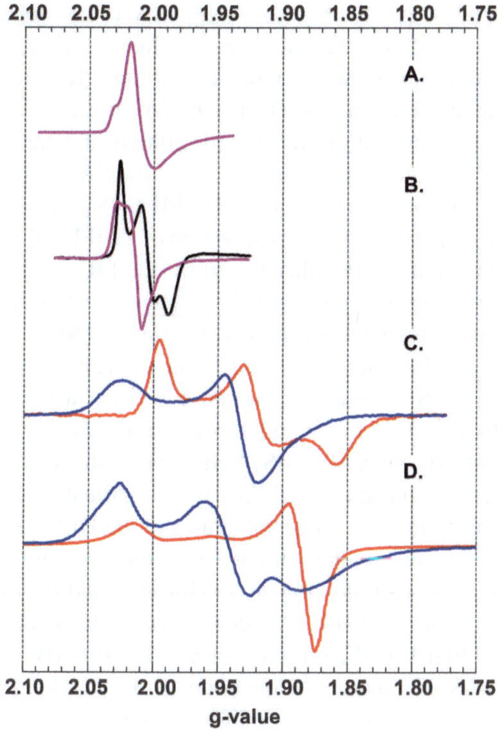

Fig. 1 EPR spectra of various states of the FeS_{RS} cluster. (*A*) $[3Fe4S]^{1+}$ (*magenta*) observed in as-isolated DesII [18]; (*B*) $[4Fe–4S]^{3+}$ (*black*) and $[3Fe–4S]^{1+}$ (*magenta*) observed in as-isolated and oxidized LAM [17]; (*C*) $[4Fe–4S]^{1+}$ in the absence (*blue*) and presence (*red*) of SAM in reduced RNR-AE [19]; (*D*) $[4Fe–4S]^{1+}$ in absence (*blue*) and presence (*red*) of SAM in reduced PFL-AE [7]. Adapted with permission from corresponding sources

cluster upon addition of iron and a reductant. Frey and coworkers [17] reported that as-isolated LAM contains a [4Fe–4S] cluster in the $[4Fe–4S]^{3+}$ state that becomes EPR silent upon reduction using dithionite paired with methylviologen as a redox mediator. Subsequent addition of ferricyanide or oxygen resulted in the appearance of an EPR signal typical of a $[3Fe–4S]^{+}$ cluster that is accompanied by an EPR signal at g = 4.3, characteristic of adventitiously bound Fe(III), indicating oxidative decomposition (Fig. 1, B). Preincubation of the oxidized enzyme with Fe(II) and addition of a reducing system resulted in recovery of approximately 90% of the activity of the pre-oxidized enzyme, indicating that the active $[4Fe–4S]_{RS}$ form could be restored.

A similar study was performed on pyruvate formate lyase-activating enzyme (PFL-AE) using Mössbauer spectroscopy [20]. The Mössbauer spectrum of anaerobically isolated PFL-AE was found to contain a mixture of $[4Fe–4S]^{2+}$, cuboidal $[3Fe–4S]^{+}$ (majority), linear $[3Fe–4S]^{+}$ and $[2Fe–2S]^{2+}$ signals, and upon reduction by dithionite, 66% of the associated iron was converted to $[4Fe–4S]^{2+}$ clusters and 12% of the associated iron was converted to $[4Fe–4S]^{1+}$ clusters, with the remaining 22% of the iron being attributed to adventitiously bound Fe(III).

In a later study, Frey and coworkers [21] showed that the [4Fe–4S] cluster in LAM can be reduced further to a $[4Fe–4S]^{1+}$ state in the presence of SAM. Moreover, addition of SAH instead of SAM resulted in a similar EPR spectrum of the

$[4Fe\text{-}4S]^{1+}$ cluster even though SAH does not support LAM-dependent turnover. It is worth noting that [4Fe–4S] clusters in the majority of metalloproteins can only access one of the two redox transitions, $[4Fe–4S]^{1+/2+}$ or $[4Fe–4S]^{2+/3+}$, at physiological redox potentials. The fact that the presence of SAM or SAH allowed for a further reduction of a cluster provides one of the earliest indications of direct interaction between the [4Fe–4S] cluster and the SAM molecule.

The redox properties of the $[4Fe–4S]_{RS}$ cluster of LAM make it an outlier in the RS superfamily. The observation of a HiPIP-like $[4Fe–4S]^{3+}$ cluster in the as-isolated state and the inability to reduce the $[4Fe–4S]^{2+}$ state to the $[4Fe–4S]^{+}$ state using typical reducing systems in absence of SAM or SAH is unusual among the characterized members of the RS superfamily. Most commonly, the $[4Fe–4S]_{RS}$ cluster is in an EPR-silent $[4Fe–4S]^{2+}$ state in the as-isolated enzyme and can be reduced to a $[4Fe–4S]^{1+}$ form in the absence of SAM with reductants such as sodium dithionite. To date, LAM is the only example among the members of RS enzyme superfamily that exhibits a $[4Fe–4S]^{3+}$ cluster in a steady state. However, it is worth noting that in the later study of LAM it was indicated that an extensive Fe/S reconstitution procedure eliminated signatures of the HiPIP-like EPR signal [22].

Several studies indicate that the binding of SAM to the $[4Fe–4S]_{RS}$ cluster increase its redox potential. Hinckley and Frey used coulometric titrations to measure the $[4Fe–4S]^{+/2+}$ redox potential in reconstituted LAM and found values of −430 mV in the presence of SAM, −460 mV in the presence of SAH, [22, 23] and −479 mV in an as-reconstituted state [24]. In PFL-AE, only a fraction of the sample can be reduced using sodium dithionite in absence of SAM, but in the presence of SAM the reduction is much closer to complete, [25] which also indicates an increase in the $[4Fe–4S]^{+/2+}$ redox potential upon SAM binding. A protein film electrochemical study of redox properties of an RS enzyme, performed by Maiocco *et al.* on BtrN, indicated an upshift of the redox potential of the $[4Fe–4S]_{RS}$ cluster by about 50 mV when SAM is present [26].

In the majority of reported cases, the effect that SAM binding has on the electronic structure of the $[4Fe–4S]_{RS}$ cluster is also evident from the shift of the principal g-values in the CW EPR spectrum (Fig. 1, C and D). In the absence of SAM, the g-values of $[4Fe–4S]_{RS}^{+}$ clusters are typically similar to the values observed for other iron-sulfur cluster-containing metalloproteins such as ferrodoxins, whereas the presence of SAM results in a much lower g_{av}, where $g_{av} = (g_x + g_y + g_z)/3$. In a few cases, all three g-values were found to be below $g_e = 2.0023$, although the exact nature of this phenomenon and its connection to other properties of the [4Fe–4S] clusters are yet to be understood. Figure 2 shows a distribution of g-values in different RS enzymes with the values presented in Table 1. Overall, these data show that binding of SAM to RS enzymes is one of the triggering events that not only initiates a reaction, but also primes the metallocofactor for the reactivity needed.

The first evidence of SAM binding to the unique Fe site was obtained in studies of PFL-AE using ^{57}Fe Mössbauer spectroscopy by taking advantage of the lability of this Fe site. Removal of the released fourth Fe ion by size exclusion chromatography in the oxidized sample and addition of ^{57}Fe during the reductive reconstitution of the [4Fe–4S] cluster resulted in the installation of a single ^{57}Fe ion at the unique

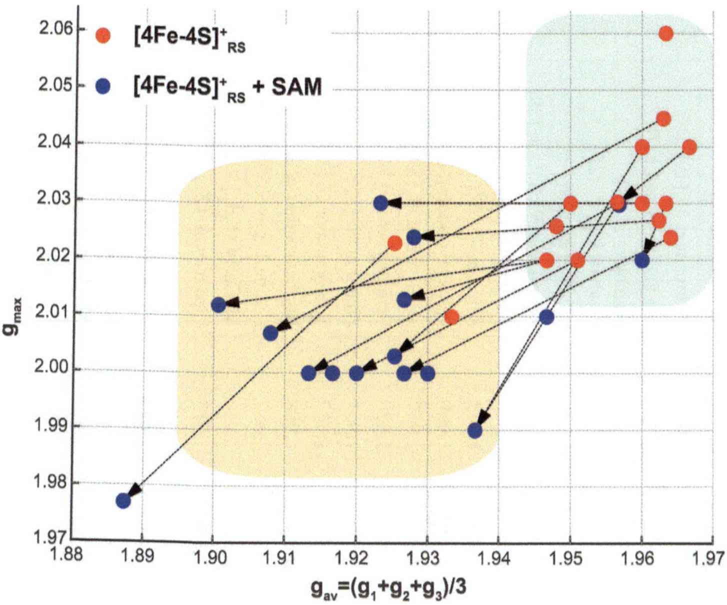

Fig. 2 Effect of SAM binding on g-values of the $[4Fe–4S]_{RS}^{1+}$ cluster in RS enzymes. Values for the corresponding data points can be found in Table 1

Fe site of the $[4Fe–4S]^{2+}$ cluster [6]. Subsequent ^{57}Fe-Mössbauer measurements revealed that adding SAM dramatically changes the ^{57}Fe isomer shift from 0.45 to 0.7 mm/s, indicating a direct binding of a strong ionic ligand. Using a reverse procedure of installing a natural-abundance Fe ion into uniformly ^{57}Fe labeled PFL-AE resulted in Mössbauer parameters for the three other iron sites that are nearly unchanged upon SAM binding. The results indicate that SAM directly binds to the unique Fe site and that the unique Fe site is electronically decoupled or valence localized. Because the measurements were performed on the $[4Fe–4S]^{2+}$ state, this study also indicated that SAM can bind to the cluster prior to the reduction event.

Hoffman and coworkers used pulse EPR techniques to further elucidate the binding mode of SAM to the unique iron site of the [4Fe–4S] cluster of PFL-AE [7, 8, 44]. Photoreduction of PFL-AE using 5-deazariboflavin afforded the $[4Fe–4S]^{+}$ cluster in high yield. The EPR spectrum changed considerably in the presence of SAM with a decrease in the g-value and a change of the spectral envelope from a very rhombic signal to an axial signal. Using SAM labeled with ^{13}C- and ^{2}H-isotopes in the methyl group, the authors performed a series of ENDOR studies using Q-band Mims ENDOR to illustrate the proximity of the SAM molecule to the [4Fe–4S] cluster (Fig. 3b, c) [8].

Broad ENDOR signals found in the ^{2}H region of the ENDOR spectrum were attributed to three ^{2}H HF couplings, suggesting a close proximity of the methyl group to the [4Fe–4S] cluster. The field dependent ENDOR measurements revealed one

Table 1 Representative g-values of the $[4Fe–4S]_{RS}^{1+}$ cluster in RS enzymes

	Without SAM				With SAM				
Enzyme	g_1	g_2	g_3	g_{av}	g_1	g_2	g_3	g_{av}	References
*Ec*RNRIII-AE	2.030	1.920	1.920	1.957	1.990	1.910	1.910	1.937	Padovani et al. [19]
*Ll*RNRIII-AE	2.024	1.934	1.934	1.964	2.000	1.920	1.860	1.927	Liu and Graslund [27]
PFL-AE	2.020	1.940	1.880	1.947	2.013	1.889	1.878	1.927	Walsby et al. [8] Broderick et al. [25]
*Ca*SPL	2.030	1.930	1.920	1.960	2.030	1.920	1.820	1.923	Silver et al. [28]
*Ac*SPL	2.027	1.930	1.930	1.962	2.024	1.930	1.830	1.928	Rebeil and Nicholson [29]
SPL_Ac	2.026	1.928	1.890	1.948	n.r.	n.r.	n.r.	–	Yang et al. [30]
ThiGH	2.030	1.920	1.920	1.957	2.000	1.870	1.870	1.913	Kriek et al. [31]
Elp3	2.030	1.930	1.930	1.963	2.020	1.930	1.930	1.960	Paraskevopoulou et al. [32]
HmdB	2.040	1.930	1.930	1.967	2.030	1.920	1.920	1.957	McGlynn et al. [33]
NirJ	2.020	1.930	1.903	1.951	2.000	1.897	1.863	1.920	Brindley wt al. [34]
NocL	2.020	1.910	1.910	1.947	2.012	1.892	1.798	1.901	Zhang et al. [35]
BtrN	2.040	1.920	1.920	1.960	1.990	1.990	1.830	1.937	Yokoyama et al. [36]
HydG	2.045	1.936	1.908	1.963	2.007	1.878	1.839	1.908	Kuchenreuther et al. [37]
TYW1	2.023	1.895	1.858	1.925	1.977	1.858	1.827	1.887	Perche-Letuvee et al. [38]
BlsE	n.r.	n.r.	n.r.	–	2.000	1.930	1.860	1.930	McCarty et al. [39]
LAM	n.r.	n.r.	n.r.	–	2.000	1.900	1.850	1.917	Lieder et al. [21]
HemN	2.060	1.940	1.890	1.963	n.r.	n.r.	n.r.	–	Layer et al. [40]
DesII	n.r.	n.r.	n.r.	–	2.010	1.960	1.870	1.947	Szu et al. [41]
PhnJ	2.010	1.920	1.870	1.933	n.r.	n.r.	n.r.	–	Kamat et al. [42]
QueE	2.030	1.910	1.910	1.950	2.003	1.912	1.861	1.925	McCarty et al. [39]
TsrM	2.020	1.922	1.860	1.934	1.975	1.936	1.901	1.937	Blaszcyk et al. [43]

n.r. not reported

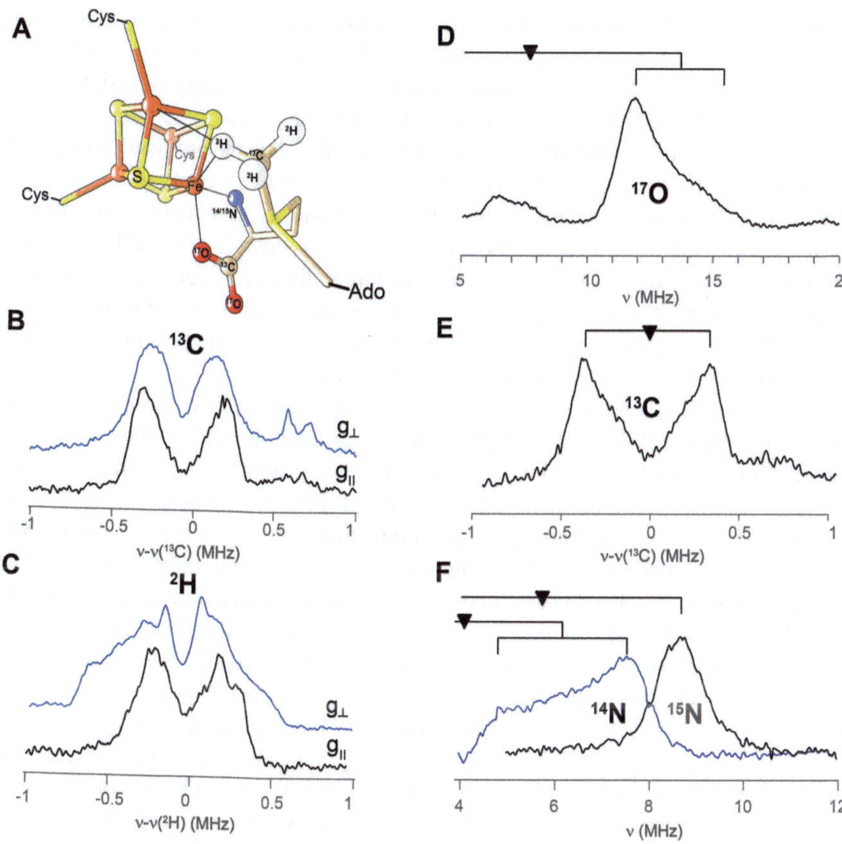

Fig. 3 (**a**) Schematic representation of SAM binding to the $[4Fe–4S]_{RS}$ cluster based on the X-ray crystal structure of PFL-AE (PDB: 3CB8). (**b, c**) Q-band ENDOR spectra of the $[4Fe–4S]^{1+}_{RS}$ cluster in the presence of ^{13}C- and ^{2}H-Methyl-SAM, respectively, adapted with permission from [8]; (**d–f**) ENDOR spectra of the $[4Fe–4S]^{1+}_{RS}$ cluster in the presence of ^{17}O-, ^{13}C-, and ^{15}N-labeled SAM, adapted with permission from [7, 44]

stronger and two weaker ^{2}H HF coupling constants with substantial anisotropy. More details about the relative positioning of SAM and the $[4Fe–4S]^{1+}$ cluster were obtained from field-dependent ^{13}C ENDOR measurements. The in-depth analysis of the obtained data accounted for the ESE signal intensity and the periodic τ-dependent blind-spot of the Mims ENDOR. The following principal values for the ^{13}C HF coupling were extracted: $A(^{13}C) = [-0.6, 0.4, -0.5]$ MHz. The slight rhombicity and the presence of the isotropic component led the authors to conclude that a small amount of spin density must reside directly on the methyl group. It was also noted that the opposite sign of the local (due to spin density on the methyl group) and non-local (through space) contributions to $A(^{13}C)$ indicate that the spin density on the methyl group does not originate directly from the [4Fe–4S] cluster via spin delocalization but rather is a result of spin polarization through an intervening atom, e.g. the sulfonium atom of SAM. Because of the $1/R^3$ dependence of the dipolar coupling constants on the distance (R), the authors concluded that only the distance to the nearest Fe site needs to be taken into account. Considering the uncertainty of assigning the dipolar interaction to a specific iron site of the [4Fe–4S] cluster and thus the uncertainty in the choice of the spin-projection factor, the authors provided a range of distances for $R(\text{Fe–}^2\text{H}) = 3.0–3.8$ Å and $R(\text{Fe–}^{13}\text{C}) = 4–5$ Å. To verify that SAM does not change its position relative to the [4Fe–4S] cluster upon reduction, the authors performed an experiment in which the $[4Fe–4S]^{2+}$ was cryoreduced at 77 K by exposing the frozen sample to a ^{60}Co source. This procedure allows for reduction of the Fe/S cluster in a frozen matrix and prevents any major structural rearrangements. The ^{2}H and ^{13}C ENDOR signals obtained for such samples were found to be almost identical to the ones obtained for the solution-reduced enzyme.

Later, Hoffman and coworkers extended the study of SAM binding to the $[4Fe–4S]_{RS}^{+}$ cluster of PFL-AE using specific isotope labeling of the carboxyl group of the methionine moiety of SAM with ^{17}O and ^{13}C and the amino group with ^{15}N [7, 44]. Q-band ENDOR measurements of the reduced PFL-AE sample prepared with ^{17}O-labeled SAM exhibited strong and predominantly isotropic ^{17}O HF coupling ([8.6, 14.4, 8.2] MHz) [44]. The overall magnitude of the ^{17}O HF coupling constant was similar to that of the carboxylate-^{17}O of citrate coordinated to the unique Fe in aconitase [45]. Orientation selective Q-band ENDOR measurements of carboxylate-^{13}C – SAM revealed distinct ^{13}C signals ($A(^{13}C) = [1.1, –0.95, –0.82]$ MHz). Once again, the authors decomposed the obtained ^{13}C coupling constant into two components, non-local ($T_{non\text{-}loc} = [-1.28, 0.64, 0.64]$ MHz) and local interactions ($T_{loc} = [-0.04, 0.09, -0.04]/a_{loc} = 0.22$ MHz) , i.e. $A(^{13}C) = T_{non\text{-}loc} + (T_{loc} + a_{loc})$. In this case, the authors used a known spin-projection factor of 1.57 for the labile Fe site of the $[4Fe–4S]^{1+}$ cluster of aconitase to provide an estimate of the distance between the unique Fe site and the ^{13}C group $R(\text{Fe-}^{13}\text{C}) = 3.3$ Å. Q-band ENDOR experiments using amino-^{15}N-SAM showed the presence of a strongly coupled ^{14}N/^{15}N nucleus in the vicinity of the $[4Fe–4S]_{RS}$ cluster. Analysis of the orientation dependence of the ^{15}N ENDOR signal yielded the following principal ^{15}N HF coupling constants: $A(^{15}N) = [9.7, 6, 3.5]$ MHz. This coupling was found to be very similar to those typically found for histidines coordinated to the Fe(II) ion of the Rieske $[2Fe–2S]^+$ cluster, thus indicating the direct binding of the amine to the unique Fe site in PFL-AE. From an analysis of all of their observations, the authors developed a SAM binding model, which was later verified by X-ray crystallographic data.

The metrics of SAM positioning with respect to the [4Fe–4S] cluster provided a hint about the mechanism of reductive homolytic cleavage of the S–C5′ bond. The fact that the sulfonium of SAM is in close proximity to the unique iron ion of the cluster suggests the possibility of a direct orbital overlap between the two. Consequently, this arrangement would allow for an efficient inner-sphere electron transfer from the $[4Fe–4S]^+$ cluster to SAM to initiate the homolytic S–C5′ bond cleavage.

HYSCORE spectroscopy has proven to be another useful technique for observing the bidentate binding of SAM to the $[4Fe–4S]_{RS}$ cluster. This technique takes advantage of the fact that in the absence of SAM, the HYSCORE spectrum of the $[4Fe–4S]_{RS}^{+}$ cluster contains only a minor signal (~4 MHz at X-band) resulting from "matrix" nitrogen atoms, likely from the peptide backbone of the cysteine ligands.

A single HYSCORE measurement taken at the maximum absorption of the $[4Fe–4S]^{+}_{RS}$ signal (at g = 1.92–1.93) was used in a variety of systems to identify such a binding mode. The first (to our knowledge) experiment of such a kind on a RS enzyme was performed on the activating enzyme of the class III ribonucleotide reductase (RNR) from *Escherichia coli* [46]. A broad and complex set of signals was found across the (++) and (+−) quadrants of the HYSCORE spectrum, which were absent in the sample without SAM. Using the known relation [47] between the frequencies of double quantum ridges in the (+−) quadrant of the HYSCORE spectrum and the ^{14}N HF coupling (***a***) and quadrupole (***K***) coupling constant (Eq. 1), Gambarelli et al. [46] identified an isotropic component of the ^{14}N HF coupling constant (a_N) of 6.4 MHz. Similar studies of spore photoproduct lyase and TYW1 yielded a_N values of 6.5 MHz [48] and 5.6 MHz, respectively [38].

$$\nu_{dq\pm} = 2\left[\left(\nu_I \pm a/2\right)^2 + K^2\left(3+\eta^2\right)\right]^{1/2} \tag{1}$$

A recent study of the cobalamin-dependent RS enzyme TsrM, a methyltransferase, provides the first example of an RS enzyme in which SAM does not bind to the [4Fe–4S] cluster, at least in the usual bidentate fashion [43]. The X-band HYSCORE spectrum of TsrM remained unchanged when SAM was added, although both cob(II)alamin and $[4Fe–4S]^{+}$ EPR signals were found to be sensitive to the addition of SAM. Neither the ^{14}N(amino) nor ^{13}C(carboxylate) nuclei signals could be resolved. Additionally, the authors used the class A RS methylase RlmN as a positive control, for which both signals could be clearly observed in a single X-band HYSCORE measurement (see Fig. 4). Using the equation for extraction of the a_N from above, we estimate the a_N(RlmN) to be 5.4 MHz.

In the most recent study of the reaction of PFL-AE by Hoffman, Broderick and coworkers, a new EPR species (denoted Ω) has been trapped by a rapid-freeze quench technique [49]. The signal has g-values of [2.035, 2.004, 2.004], that somewhat resemble the signals of a $[4Fe–4S]^{3+}$ cluster with the intensity maximum at 500 ms of freeze-quench-delay time (Fig. 5). When annealed at temperatures up to 220 K, the EPR signal of Ω is replaced with a signal of a glycyl radical, indicating that this EPR active species precludes the hydrogen abstraction step. To verify the nature of intermediate Ω, the authors performed a series of ENDOR measurements using [adenosyl-$^{13}C_{10}$]- and $^{13}CH_3$-SAM and ^{57}Fe-labeled PFL-AE. When [adenosyl-$^{13}C_{10}$]-SAM was used to prepare this species, a strong ^{13}C(adenosyl) coupling was observed ($a_{iso}(^{13}C)$ = 9.4 MHz , $T(^{13}C)$ = 2.65 MHz). In contrast, when $^{13}CH_3$-SAM was used, the corresponding ^{13}C HF coupling constant was found to be about 20 times smaller than the one from the adenosyl moiety (~0.5 MHz). Moreover, ^{57}Fe-isotope labeling of PFL-AE resulted in a ^{57}Fe HF coupling of $A(^{57}Fe)$~34 MHz that is comparable to the one observed for a $[4^{57}Fe–4S]^{1+}$ cluster prepared without SAM. Overall, the results indicated that the observed intermediate signal results from both the [4Fe–4S] cluster and the adenosyl part of SAM. The magnitude of the ^{57}Fe HF coupling indicates that the spin density of this EPR-active species resides on the Fe/S cluster, the adenosine is bound to the unique iron site by the C5′ of the adenosine (i.e. the strong

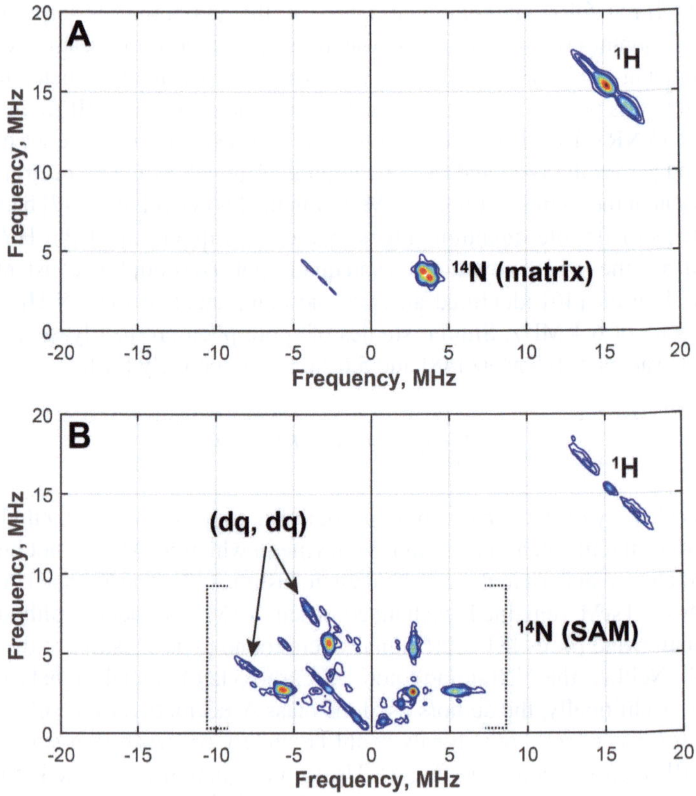

Fig. 4 X-band HYSCORE spectra of the reduced RlmN in the absence (**a**) and the presence (**b**) of SAM [43]

^{13}C HF interaction observed is due to C5′), and SAM is cleaved such that the terminal methyl group of ^{13}C-labeled methionine does not interact strongly with the EPR-active species. Overall, this conclusion points to a $[4Fe–4S]^{3+}$-[5′-dA] configuration of the intermediate. Two alternative scenarios for the formation of this intermediate have been proposed by the authors: (1) the 5′-dA• formed by reductive cleavage of SAM adds to the Fe/S cluster; or (2) a nucleophilic attack of the Fe_a site of the $[4Fe–4S]^+$ cluster on C5′ results in a formation of a Fe–C5′ bond and release of methionine. The authors propose that the Fe–C5′ bond is subsequently homolytically cleaved, resulting in a formation of the 5′-dA• and a $[4Fe–4S]^{2+}$ cluster.

Auxiliary Cofactors

In many cases, RS enzymes contain more than one Fe/S cluster or another metallocofactor. Most commonly, the auxiliary cofactor is another [4Fe–4S] cluster or a series of [4Fe–4S] clusters [50, 51]. Other Fe/S cluster forms, including a [2Fe–2S]

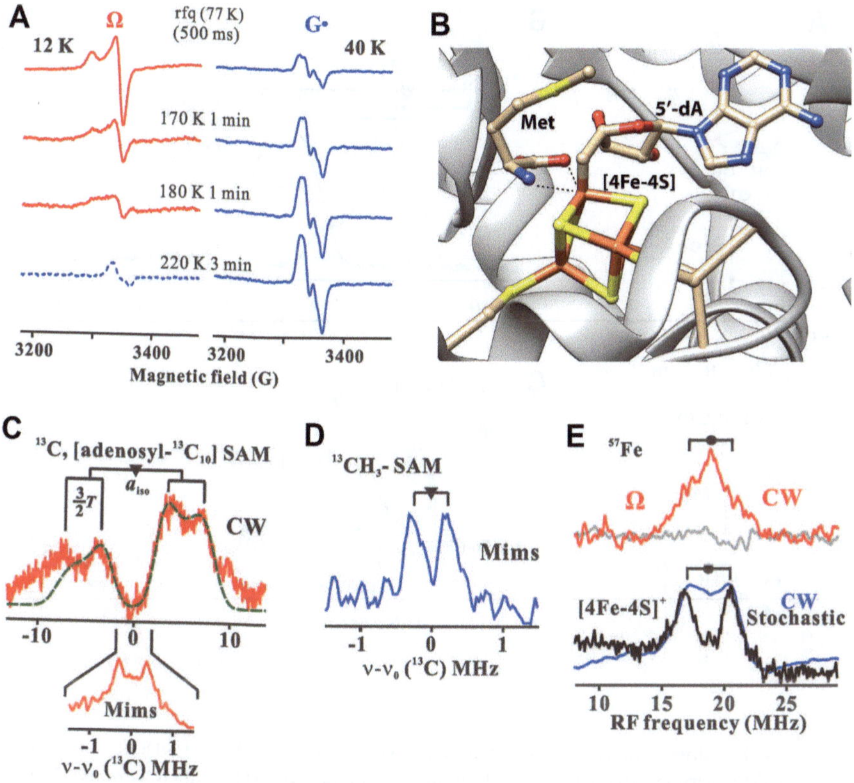

Fig. 5 (**a**) X-band EPR spectra of PFL-AE indicating decay of the intermediate Ω and formation of the glycyl radical (G•) upon annealing at different temperatures. (**b**) Structural model of the intermediate Ω (**c**–**e**) Q-band ENDOR spectra of the intermediate Ω, prepared using [adenosyl-$^{13}C_{10}$] (**c**), $^{13}CH_3$-SAM (**d**) and ^{57}Fe-labeled PFL-AE (**e**). Adapted with permission from Horitani et al. [49]

cluster, can also be found in RS enzymes. Another subfamily of radical SAM enzymes requires a cobalamin cofactor for catalysis. In many reported cases, the auxiliary metallocenters act as active components of the reaction rather than as simple mediators of electron transfer; however, much still remains to be learned about their exact roles in many RS enzymes. Advanced EPR methods, however, have played an important role in deciphering the identity and the function of several auxiliary metallocenters in RS enzymes.

[4Fe–4S]$_{aux}$ as a Binding Site

MoaA is an RS enzyme that is involved in the biosynthesis of the biologically important molybdenum cofactor (Moco), and operates in concert with MoaC to convert guanosine triphosphate (GTP) to 1,1′-dihydroxy-2′,4′-cyclic pyranopterin monophosphate (cPMP).

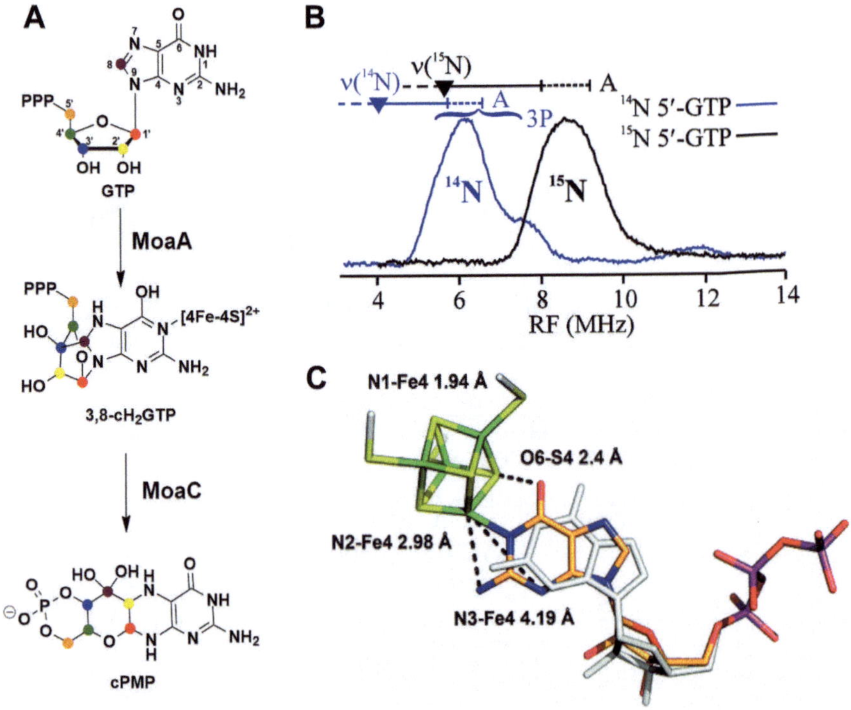

Fig. 6 (**a**) Transformation catalyzed by MoaA and MoaC. (**b**) Q-band CW ENDOR spectra of the $1(^{14,15}N)$–MoaA triple variant at 2 K. (**c**) Proposed model for 5′-GTP binding (C, *orange*; N, *blue*; O, *red*; P, *purple*) to the Fe4 ion of $[4Fe–4S]_{aux}$ (S, *yellow*; Fe, *green*). The 5′-GTP model determined by X-ray crystallography (PDB entry 2FB3) is shown in white. Adapted with permission from Lees et al. [52]

MoaA contains two [4Fe–4S] clusters that are each ligated by three cysteines. The N-terminal cluster is coordinated by cysteines in the canonical CX_3CX_2C RS motif and facilitates the cleavage of SAM, while the auxiliary C-terminal cluster is ligated by cysteines in a $CX_2CX_{13}C$ motif. The mode of GTP binding to the C-terminal $[4Fe–4S]_{aux}$ cluster was elucidated using Q-band CW ENDOR spectroscopy in concert with a C24S/C28S/C31S variant, which lacks the $[4Fe–4S]_{RS}$ cluster. This triple variant of MoaA allowed the authors to study the auxiliary cluster by pulse EPR in the absence of signals from the $[4Fe–4S]_{RS}$ cluster, which could complicate the analysis. ENDOR measurements on the S = 1/2 signal of the $[4Fe–4S]^{+}_{aux}$ cluster of MoaA revealed the presence of a strongly coupled ^{14}N nucleus, suggesting that a nitrogen atom of GTP binds to the cluster. The signal assignment was confirmed using uniformly ^{15}N-labeled GTP, which shifted the signal by approximately 3 MHz (Fig. 6). Extensive orientation-dependent measurements were performed to extract the complete set of HF coupling constants. The resulting simulation yielded

$A(^{14}N)$ = [2.4, 5.6, 4.1] MHz and $A(^{15}N)$ = [3.4, 7.8, 5.7]. To determine the exact binding orientation, the authors performed a similar study using a substrate analogue, inosine 5′-triphosphate (ITP), which differs from GTP by the absence of the exocyclic amino group. Similar ^{14}N signals were observed in the orientation-selective ENDOR spectra of the ITP-treated MoaA with $A(^{14}N)$ = [0.0, 5.1, 4.0] MHz. Using pulse Q-band Mims ENDOR on the cysteine triple variant of MoaA in concert with ^{15}N-labeled GTP, the authors were able to identify two other neighboring ^{15}N nuclei with $A(^{15}N)$ ~0.5 and ~0.2 MHz that were attributed to the $N2H_2$ ($R_{Fe\text{-}N} \approx$ 2.6–5.1 Å) and N3 ($R_{Fe\text{-}N}$ > 3.6 Å) sites, respectively.

Based on the data obtained, it was concluded that the auxiliary cluster in MoaA binds GTP via N1. The distances obtained allowed the authors to conclude that guanine binds as the enol tautomer rather than the typically favored keto form. Therefore, it was suggested that the binding-induced tautomerization is important for the mechanism of the catalytic transformation. Interestingly, the obtained metrics of GTP positioning (Fig. 6) differ from the ones observed in the X-ray crystallographic study of MoaA substrate binding, in which no direct binding was resolved. However, the authors argued that the procedures used to prepare MoaA samples for crystallography may have altered the binding mode.

A similar role for the $[4Fe\text{–}4S]_{aux}$ cluster in two methylthiotransferases, *RimO* and *MiaB*, has been hypothesized. MiaB installs a methylthiol group at C2 on an isopentenylated adenosine located at position 37 of certain tRNAs, while RimO installs a methylthiol group at C3 on Asp89 of bacterial ribosomal protein S12. Despite the difference in substrates, specifically the difference in hybridization of the target carbon atom, these two enzymes are postulated to catalyze their reactions by similar mechanisms. The mechanism by which these enzymes operate has been debated; however, the current hypothesis is that one SAM molecule donates a methyl group to an unknown thiol-containing group, which is then added to the substrate via a radical process that is initiated by reductive cleavage of another SAM molecule. It has been shown that exogenously added CH_3S^- or CH_3Se^- can be directly transferred to the corresponding products both in the MiaB and in the RimO reactions, indicating that a methylthiol molecule may be synthesized during catalysis rather than assembled directly on the substrate. Moreover, it was suggested that the auxiliary cluster does not act as a sacrificial donor but rather is a binding site for the synthesized methylthiol group. Consistent with this observation, Landgraf *et al.* showed that when RimO and MiaB are incubated with SAM in the absence of a low-potential reductant to initiate radical chemistry, time-dependent formation of methanethiol is observed upon addition of acid to the reaction mixtures.

Using ^{77}Se-labeled CH_3Se^-, Forouhar et al. [53] demonstrated that this small molecule binds to the auxiliary cluster in both of these enzymes by observing the corresponding ^{77}Se signals in X-band HYSCORE spectra (Fig. 7). The authors performed the same experiment on both wildtype MiaB as well as a variant in which the cysteine residues that coordinate the RS cluster were replaced with alanines. Unfortunately, they did not perform a complete characterization of the ^{77}Se HF

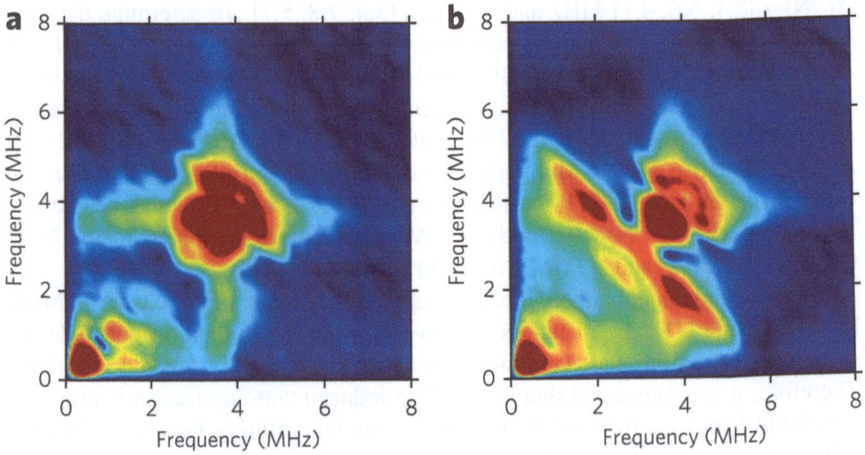

Fig. 7 X-band HYSCORE spectra of the triple cysteine to alanine substitution of the RS motif MiaB variant without (**a**) and with (**b**) $^{77}SeCH_3Na$. Adapted with permission from Forouhar et al. [53]

coupling, providing only a mean value of |A| = 3.8 ± 0.5 MHz. However, the magnitude of the HF interaction does suggest direct binding of CH_3Se^- to the [4Fe–4S] cluster. Rather interestingly, the X-ray crystallographic structure of chemically reconstituted RimO showed a polysulfide chain linking the unique iron sites of the two [4Fe–4S] clusters. Although it is most likely that such a construct is an artifact of the reconstitution process, it does indicate that the $[4Fe–4S]_{aux}$ cluster can bind exogenous sulfur atoms.

$[4Fe–4S]_{aux}$ as a Scaffold

HydG is one of the three maturation factors for [Fe-Fe] hydrogenases. It contains two Fe/S clusters: a typical N-terminal SAM-binding [4Fe–4S] cluster and an auxiliary Fe/S cluster coordinated by cysteines in a $CX_2CX_{22}C$ motif. The role of HydG is two-fold: (1) to produce CO and CN^- ligands through the breakdown of free tyrosine and (2) to form an $[Fe(CO)_x(CN)_y]$ synthon using an exogenous source of iron [37, 54, 55]. The functional aspects of tyrosine cleavage will be covered later in this chapter; here we focus on the identity and the role of the auxiliary cluster. The first reports on HydG characterized the auxiliary cluster as a [4Fe–4S] cluster [56, 57]. However, a later study of the chemically reconstituted *Shewanella oneidensis* (*So*) HydG showed the presence of an unusual S = 5/2 EPR signal (g_{eff}: 9.5, 4.7, 4.1, 3.7) that persisted when the three cysteines of the canonical RS motif were substituted with serines ($HydG^{XN}$), but was absent when the cysteines that coordinate the C-terminal cluster were altered ($HydG^{XC}$). Based on the X-ray crystallographic structure of similarly prepared wild-type *Thermoanaerobacter italicus* (*Ti*) HydG, the S = 5/2 signal was attributed to a quite unprecedented construct of a cuboid [4Fe–4S] cluster attached to a dangling, histidine-coordinated, Fe ion via a μ-sulfide bridge [58]. In addition, a labile nonproteinaceous amino acid was found near the sulfide ion. The S = 5/2 state of the auxiliary cluster in *So*HydG was explained using a simple exchange coupling model in which two spin systems $S_1([4Fe–4S]^{+1}) = 1/2$ and $S_2(Fe^{2+}) = 2$ are coupled ferromagnetically. The authors also noted that an additional observed signal at $g_{eff} = 5.5$ may originate from an S = 3/2 species due to an antiferromagnetic coupling mechanism. By simulating temperature dependent EPR signals, effective zero-field splitting parameters for the S = 5/2 species were extracted: $D_{eff} = +4.5\ cm^{-1}$, $E_{eff}/D_{eff} = 0.255$ and $g_{eff} = 2$.

The observation that tyrosine interacts with the auxiliary cluster of HydG suggests a scaffolding role for the auxiliary cluster. It was shown that addition of a reducing agent (dithionite), SAM, and tyrosine to *So*HydG results in the decay of the S = 5/2 signal and the appearance of an S = 1/2 $[4Fe–4S]^{+}_{aux}$ EPR signal (g = [2.09, 1.94, 1.93]), indicative of the loss of the dangling Fe (Fig. 8). HYSCORE measurements performed on the resulting $[4Fe–4S]^{+}_{aux}$cluster showed the presence of ^{15}N and ^{13}C signals when $^{13}C9,^{15}N$-labeled tyrosine was used, indicating that the auxiliary cluster is coordinated by a fractured derivative of the backbone part of the tyrosine [37]. Adding $K^{13}CN$ to *So*HydG resulted in a similar effect: a HYSCORE measurement of the $[4Fe–4S]_{aux}$ cluster in the presence of $K^{13}CN$ showed a clear signature of a strongly coupled ^{13}C nucleus ($A(^{13}C) = [-5.0, -4.0, 0.9]$), indicating direct binding of CN^- to the auxiliary cluster [37, 59].

Suess et al. [59] illustrated that the dangling Fe can be removed by adding a chelating agent such as EDTA, resulting in the appearance of the $g_{max} = 2.09$ signal from the S = 1/2 $[4Fe–4S]^{+}_{aux}$ cluster. Subsequent addition of exogenous Fe^{2+} results in the restoration of the S = 5/2 [5Fe] species. The observations were also confirmed by a Mössbauer study [59]. Interestingly, Ni^{2+} could be installed as the fifth metal instead of Fe^{2+}, giving rise to a high-spin signal that was attributed to an S = 3/2 state ($g_{eff} = 4.8$ and 3.6) due to ferromagnetic coupling between $S_1([4Fe–4S]) = 1/2$ and $S_2(Ni^{2+}) = 1$.

Taking into account the previous work which illustrated that an addition of nonproteinaceous cysteine increases the maturation efficiency, [55] and that an unrecognized amino acid molecule was present near the auxiliary cluster in the X-ray crystal structure, [58] it was hypothesized that a cysteine molecule constitutes the Fe–S(Cys)–Fe bridge between the [4Fe–4S] cluster and the dangling Fe. Indeed, addition of cysteine and Fe^{2+} to EDTA-treated $HydG^{XN}$ results in a complete restoration of the S = 5/2 $[5Fe]_{aux}$ species. The use of other additives instead of L- cysteine, such as D-cysteine, L-homocysteine, L-alanine plus S^{2-}, or L-serine, did not result in formation of the S = 5/2 species. To provide further evidence that cysteine binds to the $[4Fe–4S]_{aux}$ cluster, the authors performed a field-dependent ENDOR study of EDTA-treated HydG in the presence of $^{15}N,^{13}C_3$-Cys or 3-^{13}C-Cys. The orientation selective Mims ENDOR measurements of the singly ^{13}C-labeled samples revealed a pair of signals that were attributed to a ^{13}C nucleus with a roughly isotropic HF coupling (A = [0.83, 0.83, 1.09] MHz). Similar measurements of HydG prepared with $^{15}N,^{13}C_3$-Cys revealed more complex signals, but the largest signals did not

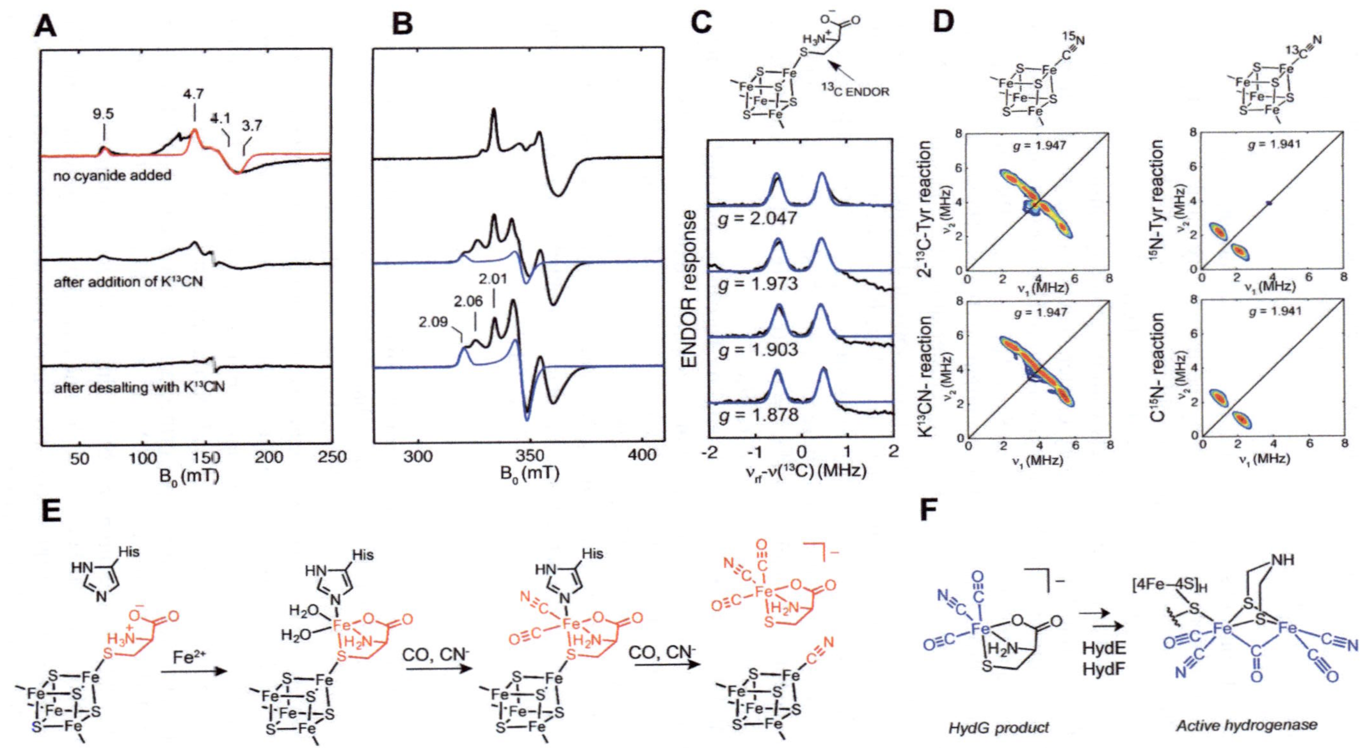

Fig. 8 Low-field (**a**) and high field (**b**) EPR signals of HydG, experimental spectra in black, where the red trace is the simulation of the S = 5/2 species and the blue trace is the simulations of $[4Fe\text{-}4S]^{+}$-Cys species, (**c**) Q-band Mims ENDOR spectra of HydG in $[4Fe\text{–}4S]^{-}$-3-^{13}C-Cys state (*black*) with simulations (*blue*) (**d**) Evidence of the presence of CN-ligand to the $[4Fe\text{–}4S]_{aux}$ cluster by ^{15}N- (*left*) and ^{13}C- (*right*) HYSCORE, (**e**) Proposed scheme of formation of an $[Fe(CO)_x(CN)_y]$ synthon on the auxiliary cluster of HydG, (**f**) Conversion of a HydG-generated synthon into the hydrogenase H-cluster by HydE and HydF. Adapted with permission from Dinis et al. [58] and Suess et al. [59]

exceed the overall width of the signals obtained for the singly labeled sample, indicating that cysteine likely binds to the $[4Fe–4S]_{aux}$ cluster via the thiol group.

Finally, Suess et al. showed that cysteine can replace cyanide at the open coordination site by observing the change in the EPR signal for the auxiliary cluster from g^{CN} = [2.09, 1.94, 1.93] to g^{Cys} = [2.06, 1.90, 1.87] when the samples were sequentially treated with KCN and Cys. Based on the obtained data, it was concluded that the auxiliary cluster is a scaffold for the formation of the $[Fe(CN)_2CO]$ synthon that is mediated by the presence of nonproteinaceous cysteine.

$[2Fe–2S]_{aux}$ as a Sacrificial Donor

Biotin synthase (BioB) is an RS enzyme that converts dethiobiotin (DTB) to biotin. Through extensive studies using a variety of biophysical methods, it was concluded that BioB acts as a dimer and contains one $[4Fe–4S]_{RS}$ and one $[2Fe–2S]_{aux}$ cluster per monomer [60, 61]. The stoichiometry and species of Fe/S clusters was later confirmed by X-ray crystallography [62].

The [2Fe–2S] cluster, coordinated by three cysteine and one arginine ligand, was shown to donate one of its μ-sulfide bridges to the substrate during turnover (Fig. 9) [63–65]. An important mechanistic insight was obtained by X-ray crystallography [62]. The structure of BioB in the presence of SAM and dethiobiotin revealed that dethiobiotin is located between the $[4Fe–4S]_{RS}$ and $[2Fe–2S]_{aux}$ clusters. C9 of dethiobiotin was located essentially in the middle between C5′ of SAM and one of the bridging thiols. Therefore, it was proposed that after the 5′-dA• abstracts a hydrogen atom from the C9 position, the substrate radical that is formed subsequently recombines with the thiol bridge of the [2Fe–2S] cluster and thus initiates the transfer of the sulfur atom to C9.

Although no substrate radical intermediates have been identified, it was shown that the [2Fe–2S] cluster does get transiently reduced under turnover conditions with kinetics that correlate with the formation of a 9-mercaptodethiobiotin (MDBT) intermediate [65, 66]. The reduced $[2Fe–2S]^+$ cluster was found to exhibit two distinct signals (g = [2.010, 1.955, 1.880] and g = [2.000, 1.940, 1.845]) that both have the same kinetic behavior, suggesting the possibility of structural heterogeneity. The identity of the paramagnetic species was verified by observation of nitrogen signals in ESEEM and HYSCORE that are characteristic of arginine ligation [65, 67]. Most notably, when BioB was expressed in the presence of *guanidino*-$^{15}N_2$–Arg, Fugate et al. [67] were able to resolve two ^{15}N signals from the guanidino group in the X-band HYSCORE spectrum.

To clarify the structure of the intermediate further, Fugate et al. [67] prepared BioB with (*9-methyl*-^{13}C)- dethiobiotin and performed a HYSCORE analysis of the $[2Fe–2S]^+$ intermediate. The ^{13}C HF coupling observed ([1.2, 1.2, 5.7] MHz) provided direct evidence that dethiobiotin is bound to the reduced $[2Fe–2S]^+$ cluster. Therefore, these measurements indicate that a dethiobiotin-radical is transiently stabilized as a bound adduct to the [2Fe–2S] cluster. Also, these results suggest that this Fe/S cluster is not just a sulfur donor but also is a catalytic cofactor that facili-

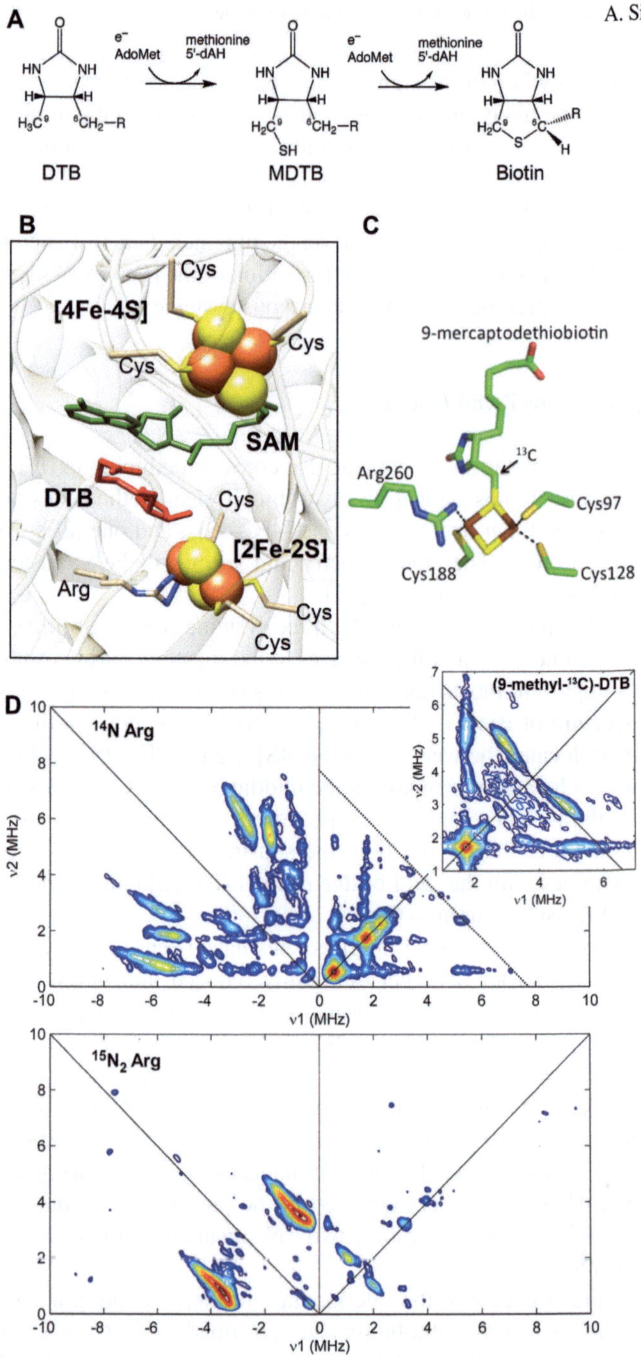

Fig. 9 (**a**) Reaction of BioB with DTB. (**b**) Relative positioning of Fe/S clusters, SAM and DTB in BioB (PDB: 1R30), (**c**) Proposed structure of an intermediate. (**d**) X-band HYSCORE spectra of the $[2Fe–2S]^+$ intermediate in BioB in natural abundance of N-isotopes (*top*) and with ^{15}N-labeled Arg (*bottom*), insert shows HYSCORE spectra obtained using (9-methyl-^{13}C)-DTB as a substrate. Adapted with permission from Fugate et al. [67]

tates inner-sphere one-electron oxidation of the sulfide concurrent with formation of the new C–S bond. This mechanism provides a means for safe termination of the MDBT radical and, therefore, allows for a subsequent hydrogen abstraction at C6 using a second 5'-dA•.

Radical Intermediates

Because the mechanisms of RS enzymes imply the presence of organic radical intermediates, great effort has been devoted to trapping and characterizing such intermediates by EPR. In the following, we describe several examples of such work.

Pyruvate Formate-Lyase Activating Enzyme (PFL-AE)

PFL-AE was one of the earliest enzymes of the RS superfamily to be studied, and in fact preceded the bioinformatic identification of the superfamily. Soon after the identification of PFL-AE as an activator of PFL, it was found to cleave SAM into methionine and 5′-dA with concomitant formation of a protein-based radical on PFL [68–70].

The CW EPR spectrum of activated PFL was characteristic of a carbon-based radical that is dominated by a single ^{1}H hyperfine splitting of 1.5 mT (Fig. 10) [71]. The doublet-shaped spectrum collapsed into a single line upon exchange into buffer prepared with D_2O, indicating that the corresponding proton is solvent exchangeable [71, 72]. The spectrum also contained a fine structure that was not affected by H_2O/D_2O exchange but was diminished when PFL was expressed in deuteurated media indicating that this feature is due to non-exchangeable protons. The final identification was made by observing EPR line broadening upon selective labeling of glycine residues in PFL with either [2-^{13}C]glycine or [1-^{13}C]glycine [72]. Extracted spin-Hamiltonian parameters indicated that the spin density is localized on C2 and supported the notion that PFL-AE abstracts the C_α–H, *pro-S* H-atom of G734. These results, and the fact that only the reduced [4Fe–4S]$^+$ state is the active form, illustrated that RS enzymes utilize reductive cleavage of SAM and use the 5′-dA• generated to abstract an H-atom from a (co)substrate; which, along with the early study of lysine 2,3-aminomutase, were the first indications of the key role SAM plays in the reaction of RS enzymes.

Lysine 2,3-Aminomutase (LAM)

LAM is perhaps the best spectroscopically characterized RS enzyme. LAM catalyzes the conversion of L-α-lysine to L-β-lysine [73]. In addition to SAM and a [4Fe–4S]$_{RS}$ cluster, its reaction also requires pyridoxal 5′-phosphate (PLP), which

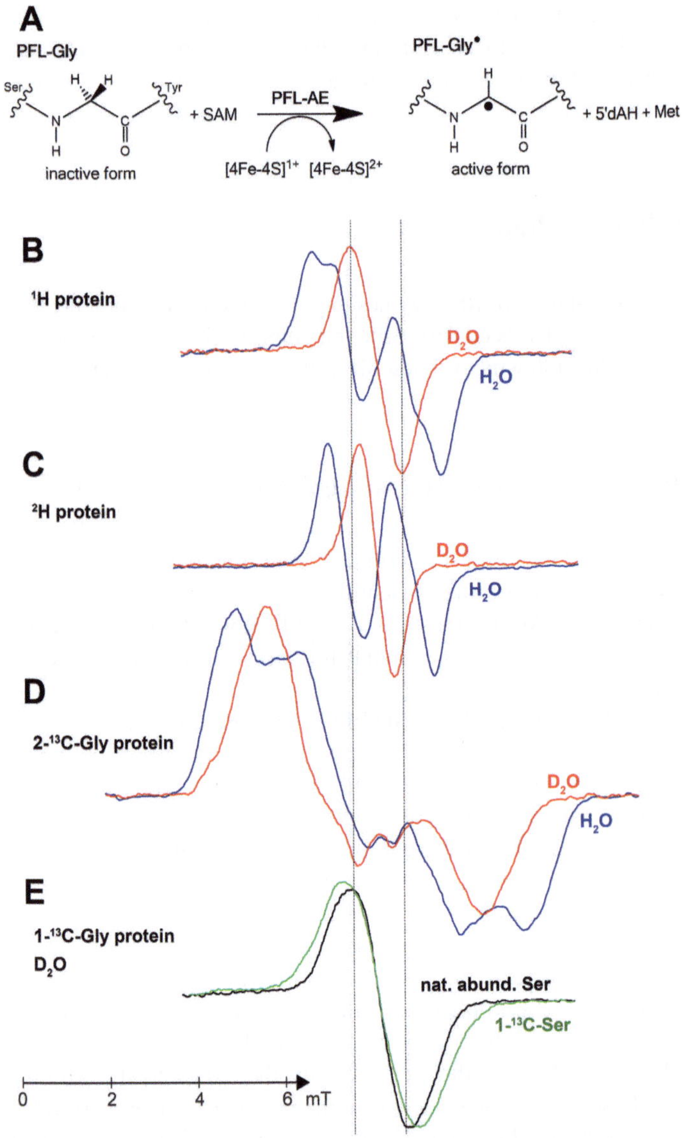

Fig. 10 Generation of PFL-Gly• by PFL-AE (**a**). X-band EPR spectra of PFL-Gly• as generated by PFL-AE for (**b**) PFL in H_2O (*blue*) and D_2O (*red*); (**c**) ^{2}H-labeled PFL in H_2O (*blue*) and D_2O (*red*); (**d**) [2-^{13}C]-Gly enriched PFL; (**e**) [1-^{13}C]-Gly enriched PFL in D_2O with unlabeled serine (*black*) and [1-^{13}C]-Ser (*green*). Adapted with permission from Unkrig et al. [71]

forms an external aldimine with the lysine substrate or β-lysine product. In the proposed reaction mechanism (Scheme 2), the SAM-derived 5′-dA• (1) initiates substrate-based catalysis by abstracting the 3-pro-*R* H-atom of lysine to yield a substrate radical intermediate (2), which undergoes formation of an

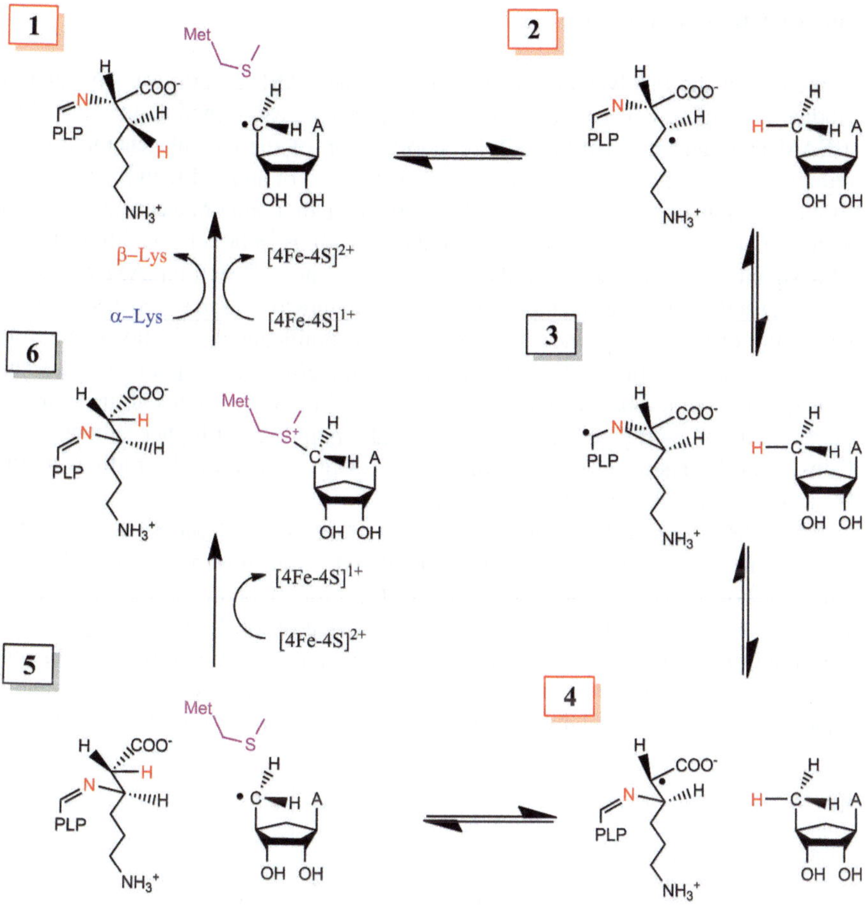

Scheme 2 Reaction of Lysine 2,3-aminomutase (LAM)

azacylopropylcarbinyl radical (3). This quasi-symmetric radical intermediate can partition backwards to form the initial substrate radical or forward to form the product radical (4). The product radical is quenched upon abstraction of an H-atom from 5′-dA to regenerate the 5′-dA•, which, in concert with the $[4Fe\text{–}4S]_{RS}$ cluster (5), recombines with methionine to regenerate SAM (6). Catalysis is completed upon transimination to form an internal aldimine with a conserved lysyl residue. In the catalytic reaction of LAM, 5′-dA formation is not stoichiometric with product formation. SAM acts as a true cofactor rather than as a co-substrate, because it is regenerated after each turnover. The mechanism shown in Scheme 2 is supported by a number of EPR studies that provided evidence for various intermediates in the reaction.

Identification of State "1"

Evidence for the involvement of the 5′-dA• in the LAM reaction was obtained in studies by Magnusson et al. using *S*-3′,4′-anhydroadenosyl-L-methionine (anSAM), a SAM analogue that permits allylic stabilization of the radical generated upon reductive cleavage of anSAM [74, 75]. Upon reaction of anSAM with LAM, a slow formation of an organic radical was observed with a first order rate constant (k_{obs}) of approximately 10 min^{-1}. At 77 K, the radical exhibited a broad, roughly isotropic EPR signal with a fine structure that disappeared when [5′-$^2H_2$]-anSAM was used. Later, the authors reported that lowering the temperature to 4.2 K resulted in significant narrowing of the unitary lineshape, leading to a significant clarification of the fine structure to distinct lines. The unusual temperature effect was explained by a weak spin-spin interaction between a radical and low-lying excited states of the formally diamagnetic $[4Fe–4S]^{2+}$ cluster that are partially populated at higher temperatures. Consequently, such an explanation implies a proximity of the radical to the $[4Fe–4S]^{2+}$ cluster upon reductive cleavage.

The EPR spectrum broadened when the [1′,2′,3′,4′,5′-^{13}C]-anSAM isotopolog was used to generate the radical, confirming that the radical is indeed located on the anhydroribosyl moiety (Fig. 11). Selective ^{2}H isotope labeling of the molecule at C5′, C3′, and C2′ resulted in a significant perturbation of the EPR lineshape and permitted the authors to extract HF coupling constants for the corresponding protons, thus resolving the electronic structure of the radical species. Using McConnel's relation for spin-polarized alpha carbons, $a^H = \rho Q$, in which Q is an empirical con-

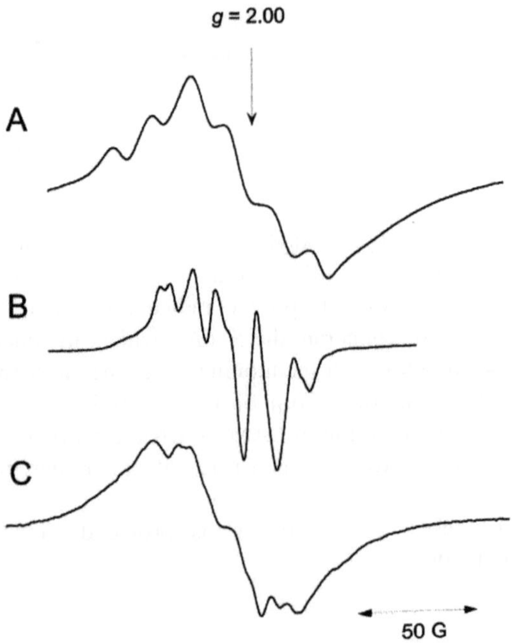

Fig. 11 Effect of temperature and ^{13}C substitution on the X-band CW EPR spectrum of the anhydroadenosyl radical (5′-*and*A•): (**a**) using unlabeled *an*SAM at 77 K. (**b**) using unlabeled *an*SAM at 4.2 K. and (**c**) using [1′,2′,3′,4′,5′-$^{13}C_5$]*an* SAM at 4.2 K. Adapted with permission from Magnusson et al. [75]

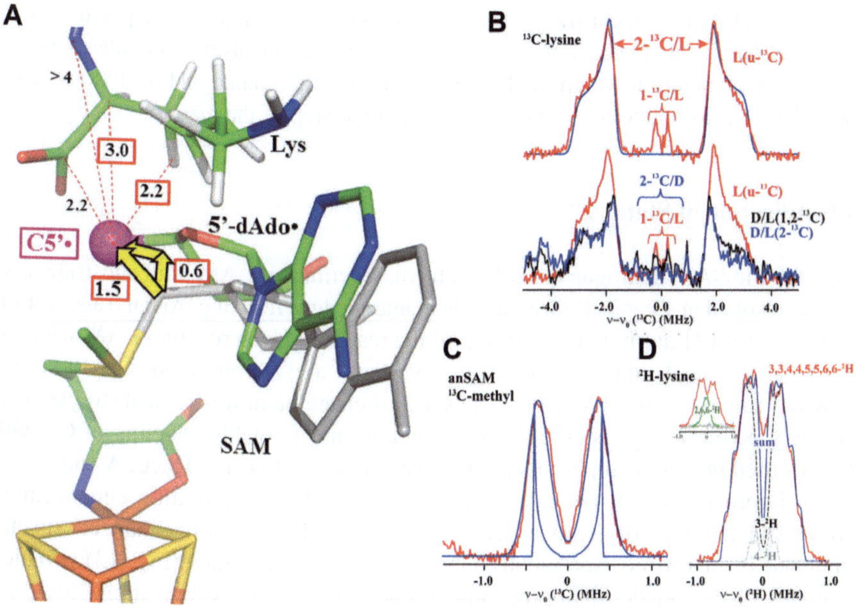

Fig. 12 (**a**) Model of relative position of lysine, methionine and 5′-dA•. Position of SAM prior reductive cleavage is shown in gray (**b**, **c**) Q-band ^{13}C ENDOR spectra of 5′-*an*dA• in LAM prepared in the presence of ^{13}C-lysine (B)or with ^{13}C-methyl-*an*SAM (C) (**d**). Q-band ^{2}H ENDOR spectra of 5′-*an*dA• in the presence of ^{2}H-labeled lysine. Adapted with permisson from Horitani et al. [76]

stant and ρ is the spin density population on the respective carbon, it was determined that the C3′ and C5′ carbons share near equally about 59% of the spin density, which is in a good agreement with the expected spin distribution for an allylic radical. It is worth noting that this study represents the most direct evidence for a 5′-dA• to date, as the direct observation of this radical species remains elusive due to its high reactivity.

More recently, Hoffman, Broderick, and coworkers analyzed the surrounding environment of the 5′-*an*dA• in LAM in great detail using ENDOR and specific ^{13}C and ^{2}H isotope labeling [76]. The results indicated that C5′ is approximately 3 Å away from C2 of the lysine substrate, and although no signal from the C3 ^{13}C nuclei could be conclusively resolved, ^{2}H measurements indicated that the 3-^{2}H is only 2.4 Å away—a perfect distance for hydrogen atom abstraction. Interestingly, although in this state of the reaction no direct bonding should exist between C5′ and lysine, a substantial isotropic component to the corresponding ^{13}C HF coupling of C2 was clearly evident from the corresponding ENDOR spectrum (Fig. 12). This effect was explained in terms of Pauli delocalization, which implies spin delocalization through space onto the neighboring atom when this atom is in tight van der Waals contact with the spin-bearing orbital of the allylic radical. Additionally, using ^{13}C-methyl *an*SAM, the authors were able to resolve a distance between C5′ and the methyl group of L-methionine of about 3.6 Å. Overall, this study indicates that after reductive cleavage, SAM fragments are essentially fixed in the same position with the substrate located near the C5′ site as well. Such a conclusion coincides with the overall mechanism in which the fractured parts (L-methionine and 5′-dA) are reassembled to a complete SAM molecule in the final step of the reaction.

Identification of State "2"

As with the SAM analogue described above, reactions of LAM with substrate analogues resulted in a trapped substrate radical intermediate resulting from the abstraction of the 3-*pro-R* H-atom in the next step of the reaction. Upon reacting LAM with the lysine analogue, 4-thia-L–lysine (SLys), Miller et al. [77] observed formation of a new radical species by CW EPR. The spectral envelope of this radical (SLys•) was strongly perturbed by specific isotopic labeling; most notably, labeling of C3 with ^{13}C resulted in significant broadening of the spectrum (Fig. 13, Traces A and E).

To improve spectral resolution, the authors applied a resolution enhancement (RE) data manipulation procedure [79]. Simulated resolution enhanced spectra obtained from various isotopically labeled substrate analogues (Fig. 13) revealed that 80% of the spin density of the radical was localized at the C3 position. However, as evident from the substantial ^{1}H HF couplings of the C5 hydrogens (3.5 G and 4.0 G), moderate delocalization of the radical across S4 takes place. Rather surprisingly, the g-anisotropy reported by the authors is very small, thus indicating negligible contributions of the spin-orbit coupling of the sulfur atom to the overall electronic structure, which possibly indicates that sulfur carries no noticable spin density. In addition, interpretation of the EPR spectra required an isotropic HF coupling from an ^{14}N nucleus of A_{iso} = 9.5 MHz. Resolving the ^{1}H hyperfine coupling constants from the C2 proton allowed for the extraction of dihideral angles of 77° and 17° between the plane defining the spin-carrying p-orbital on C3 and the C2–H bond and the C2–N(PLP) bond, respectively. Due to hyperconjugation effects, this result implies that a strong ^{14}N HF interaction with an α-amino group must take place, providing the assignment for the observed ^{14}N HF coupling. Overall, this detailed study of the electronic structure confirmed that the observed radical species is analogous to state 2 in Scheme 2.

Further support for the presence of a C3-based substrate radical in LAM was provided with another substrate analogue, trans-4,5-dehydrolysine, that allowed for stabilization of an allylic 4,5-dehydrolysyl radical (DHLys•), formed upon abstraction of the 3-*pro-R* hydrogen atom [78]. The EPR spectrum of the observed radical contained fine structure that was affected by isotope labeling of trans-4,5-dehydrolysine with ^{2}H. The HF coupling constants extracted from these experiments confirmed the delocalization of the spin density between C3 and C5 expected for an allylic radical. In addition, the authors labeled N6 with ^{15}N, which resulted in no change in the spectrum, suggesting that the observed triplet splitting is due to the α-amino group (N2). Similarly to that observed in the SLys• radical, this splitting originates from a hyperconjugation with unpaired spin density on C3. The $^{14}N2$ and $^{1}H6$ HF coupling constants and the fact that the $^{1}H2$ signal is below the limit of detection suggests that the H6–C6 and C2–N2 bonds lie essentially in the plane perpendicular to the spin-bearing p-orbitals on C3 and C5.

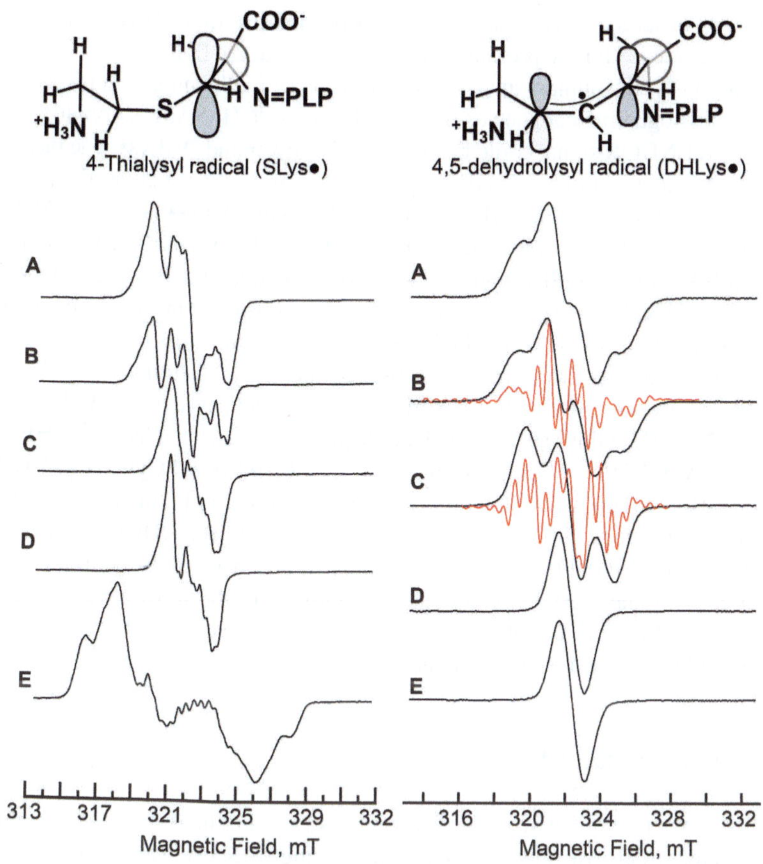

Fig. 13 *Left*: X-band CW EPR spectra of LAM with SLys bound. (*A*), 4-thialysine; (*B*), 4-thia[5,5,6,6-$^2H_4$]lysine; (*C*), 4-thia[3,3-$^2H_2$]lysine; (*D*), 4-thia[3,3,5,5,6,6-$^2H_6$]lysine; (*E*), 4-thia[3-^{13}C]lysine. Adapted with permission from Miller et al. [77]. *Right*: X-band CW EPR spectra of LAM with DHLys• bound using (*A*) unlabeled DHLys ; (*B*) [2-^{2}H]-DHLys; (*C*) [4,5-$^2H_2$]-DHLys; (*D*) [3,3,4,5,6,6-$^2H_6$]-DHLys; (*E*) [2,3,3,4,5,6,6-$^2H_7$]-DHLys. *Red spectra* in panels (*B*) and (*C*) are the results of resolution enhancement. Adapter with permission from Wu et al. [78]

The positioning of both substrate analogue radicals with respect to 5′-dA and PLP have been extensively analyzed by the Hoffman group [80]. Both the SLys• and DHLys• intermediates exhibited a relatively strong ^{13}C HF coupling with ^{13}C5′ when [5′-^{13}C]-SAM was used in the reaction (SLys•: A_{iso} = 12.7 MHz and T = 2.5 MHz; DHLys•: A_{iso} = 2.0 MHz and T = 0.6 MHz, where T and A_{iso} are constants defining the principal components of the HF interaction $A_{x,y,z} = [-1, -1, 2]T + A_{iso}$). The strong isotropic character of both HF coupling constants suggests that the 5′ carbon is in tight van der Waals contact with the radical residing on the SLys and DHLys molecules. The magnitude of the HF coupling indicates that the interaction is stronger in the case of SLys than that of DHLys. The authors refrained from further analysis of the anisotropic HF coupling, because the almost-bonding positioning of the substrate and 5′-dA is too short to directly extract the distance between the radical and C5′ of 5′-dA. In addition, the authors observed a ^{2}H signal in DHLys• originating from ^{2}H at C5′, providing additional evidence for the proximity of the 5′ carbon to the DHLys•.

Analogously, the authors analyzed the position of methionine with respect to the substrate analogue radical species in LAM. Using a $^{13}CH_3$-SAM isotopolog, a clear ^{13}C Q-band ENDOR signal could be observed for SLys• and DHLys•. The interpretation yielded an anisotropic ^{13}C HF coupling. Once again, SLys• showed a larger ^{13}C HF coupling ($T = 0.3$ MHz and $A_{iso} = 0.1$ MHz) than DHLys• ($T = 0.11$ MHz and $A_{iso} = 0.0$ MHz). Using dipolar coupling constants (T), distances from the methyl group of SAM to C3 of SLys• and DHLys• were calculated to be ~4.1 Å and 5.5 Å, respectively, which is very similar to the distances observed in the X-ray crystallographic structure of LAM with all of its substrates and cofactors bound.

Another important structural detail was obtained by detecting ^{31}P-PLP signals using the DHLys and SLys substrates. Both substrate radicals exhibited ^{31}P signals in Q-band ENDOR. In this case, the DHLys• displayed a somewhat larger ^{31}P HF tensor than the SCys• (DHLys•: $T = 0.59$ MHz and $A_{iso} = 0.19$ MHz; SCys• : $T = 0.41$ MHz and $A_{iso} = 0.19$ MHz). Using the dipolar coupling constant (T), the authors resolved $R(C3–P_{PLP})$ distances of 4.3 Å and 3.6 Å for SCys• and for DHCys•, respectively. The authors pointed out that the distance obtained is considerably smaller than what was observed in the crystal structure (7 Å) regardless of the substrate analogue considered, indicating that the phosphate group of PLP moves closer to C_β during the reaction and implying a possible isomerization of PLP (Fig. 14).

Identification of State "4"

During the reaction of LAM with its native substrate, a radical species formed in the steady state, which was absent when either lysine or SAM was omitted from the reaction. ^{2}H isotope labeling of the substrate resulted in a perturbation of the EPR lineshape, indicating that the radical species is based on the substrate or a substrate derived molecule (Fig. 15) [81].

Since the activation of LAM by SAM is much slower than the enzymatic reaction (~5 s), the time dependence of the formation of the putative product radical is not represented by the catalytic kinetic rates but rather by the activation kinetics. Therefore, it is not possible to prove the kinetic competence of this radical species directly from the kinetics of rapid-freeze quench experiments. Instead, Chang et al. [82] performed a sequential mix rapid freeze-quench experiment to illustrate that the radical species turns over as fast as the overall enzymatic turnover. In this experiment, LAM was activated using [2-^{2}H]-lysine and SAM for 5 s in the first mixing step. Then the reaction was mixed again with an excess of unlabeled lysine and freeze-quenched at various time points. If the radical is an intermediate state, the EPR spectrum of the deuterated radical species should be gradually replaced with the EPR spectrum of the unlabeled species with a rate of conversion comparable to the rate of product formation. The obtained rate of conversion of the EPR signal (24 s^{-1}) was somewhat faster than the estimated LAM turnover rate determined

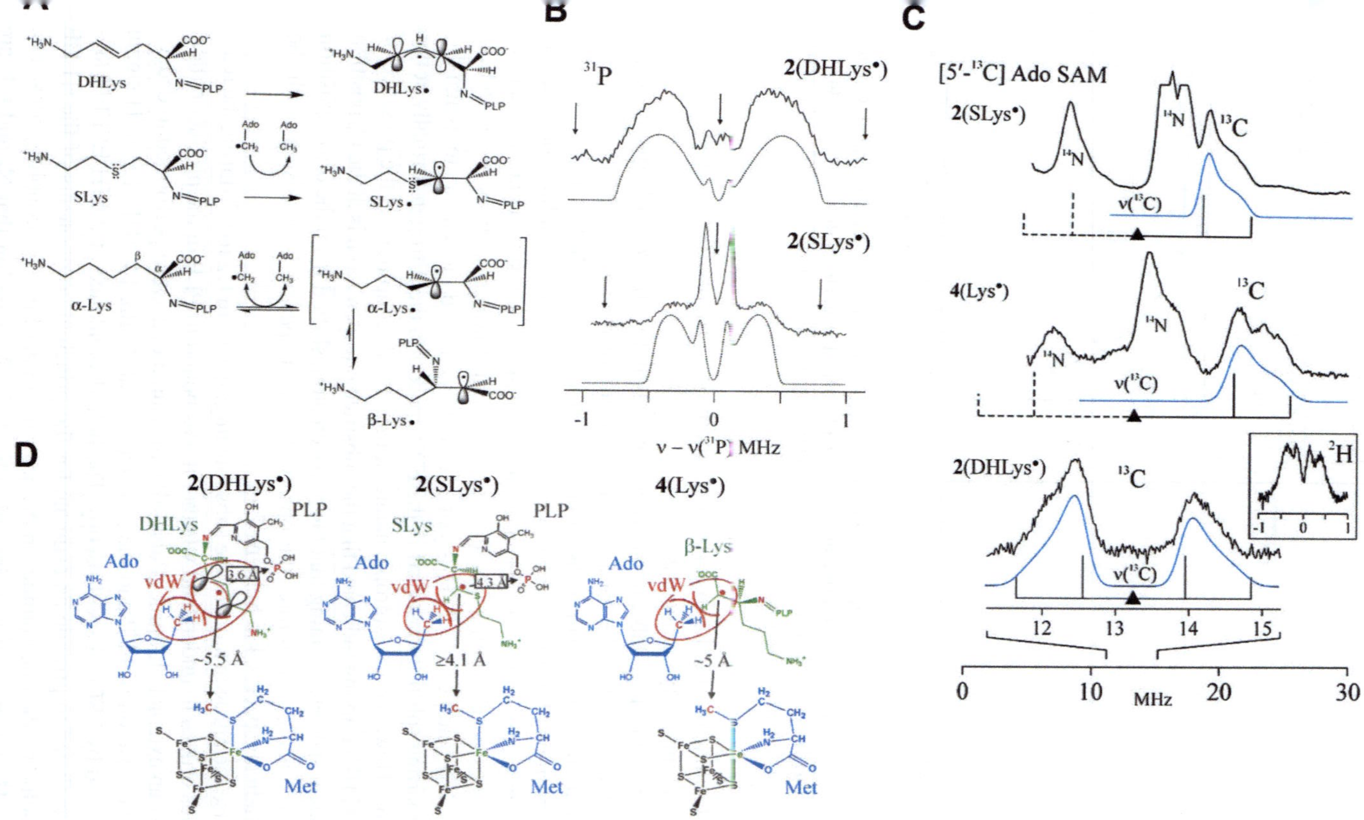

Fig. 14 (**a**). Structure of the radical species generated from substrate analogues of LAM. (**b**) ^{31}P ENDOR spectra of DHLys• and SLys•. (**c**). ^{13}C ENDOR spectra of DHLys• and SLys• and native state "4" (4(Lys•)). (**d**) Schematic representation of the interaction of DHLys•, SLys•, and 4(Lys•) with 5′-dA , L-methionine, and PLP. Note that the numbering of states was changed from the original article to comply with the numbering on the Scheme 2. Adapted with permission from Lees et al. [80]

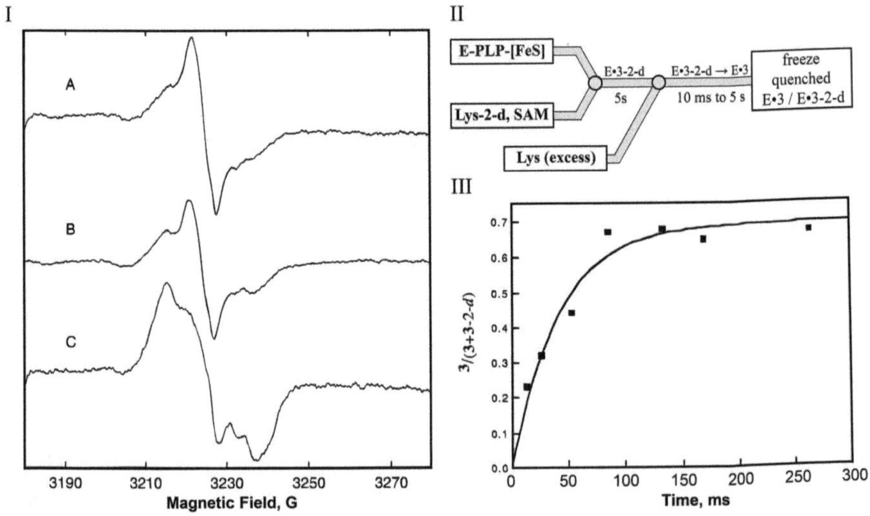

Fig. 15 Conversion of the product radical in LAM from a deuterated to a non-deuterated form in the course of a freeze quench experiment schematically depicted in (**b**) with a time dependence shown in (**c**). (**a**) depicts EPR spectra obtained by quenching the reaction at 13.6 ms (*A*), 53 ms (*B*) and 5 s (*C*) Adapted from Chang et al. [82]

from independent kinetic experiments, therefore proving the radical species observed is kinetically competent.

An extensive isotope labeling study in combination with CW EPR spectroscopy was performed to clarify the nature and the structure of the radical species. The most prominent perturbations to the EPR spectrum were achieved with ^{2}H- and ^{13}C-labeling at the C2 position of lysine, which resolved an isotropic ^{2}H HF coupling constant and a significantly anisotropic ^{13}C HF coupling constant. This result suggested that the radical resides on C2 with a single hydrogen atom. Further analysis of the radical using [2-$^{15}N_1$]- and [2-$^{15}N_1$, 2-$^{2}H_2$]-lysine substrates allowed the authors to assign the triplet hyperfine splitting observed in the EPR spectra to the nitrogen formally at the C_α position. However, the isotropic nature of the ^{14}N HF coupling and the presence of a C2–H ^{1}H HF coupling indicated that the substrate nitrogen atom is at the C_β position in the radical species, meaning that the nitrogen atom shifted from the α to β position during the reaction. Therefore, the observed radical species is consistent with the product radical (species 4 in Scheme 2).

Because the nitrogen atom of the substrate migrates from C2 to C3 during the reaction, it was crucial to understand whether it was bound to PLP in the product radical species as an external aldimine. Pulse EPR measurements were performed to deduce the relative position of PLP and the product radical. Using [4′-^{2}H] PLP, a ^{2}H signal was observed by ESEEM spectroscopy. The field dependence of the ^{2}H ESEEM signal was analyzed to provide an estimate for the principal components of the ^{2}H HF tensor. Including the uncertainties in the determination of the HF coupling constants, the R(C2–H4′) distance was estimated to be 3–3.5 Å, meaning that N_β and C4′ are within direct bonding distance [84]. Therefore, these results support the hypothesis that the PLP coenzyme facilitates radical-mediated rearrangements. More details

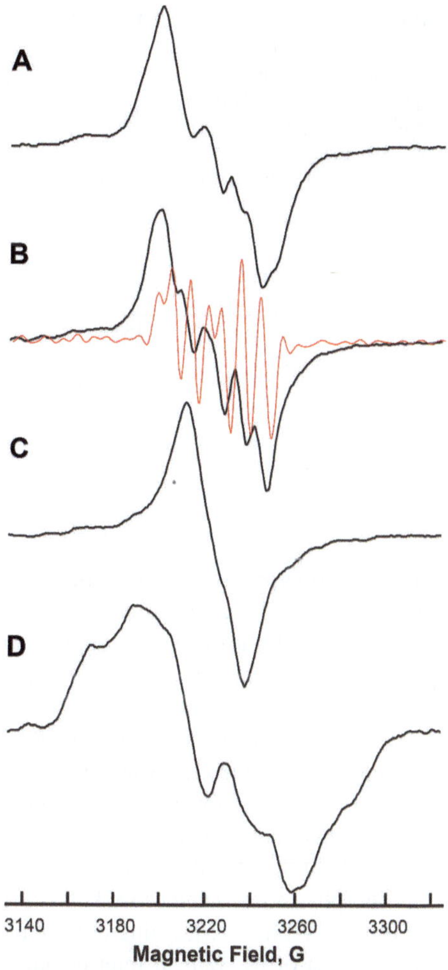

Fig. 16 The effect of isotope labeling on the X-band CW EPR spectra of the product radical. (*A*) unlabeled lysine, (*B*) L-[3,3,4,4,5,5,6,6-$^2H_8$] lysine (*C*) DL-[2-^{2}H] lysine, and (*D*) L-[2-^{13}C] lysine. The red spectrum in B is the result of resolution enhancement. Adapted with permission from Ballinger et al. [83]

about the overall positioning of the radical have been obtained through a series of ENDOR measurements analogous to the ones performed for the SLys and DHLys substrates (see above). Interestingly, the geometric metrics obtained were found to be very similar to the ones of SLys (Fig. 16) [80], indicating that despite the migration of the N-PLP moiety from C_α to C_β, the substrate remains in the same place within the pocket. This once again illustrates the tight control that the protein environment imposes on the position of the (co)substrate molecules in this class of enzymes.

HemN

An anaerobic pathway for the biosynthesis of porphyrins requires oxidative decarboxylation of the propionate side chains on the A and B pyrrole rings of coproporphyrinogen III to form the corresponding vinyl groups (see Scheme 3). This reaction is carried

Scheme 3 (**a**) Decarboxylation of the propionate side chains of coproporphyrinogen III by HemN. (**b**) Proposed substrate radical intermediate of HemN [85]

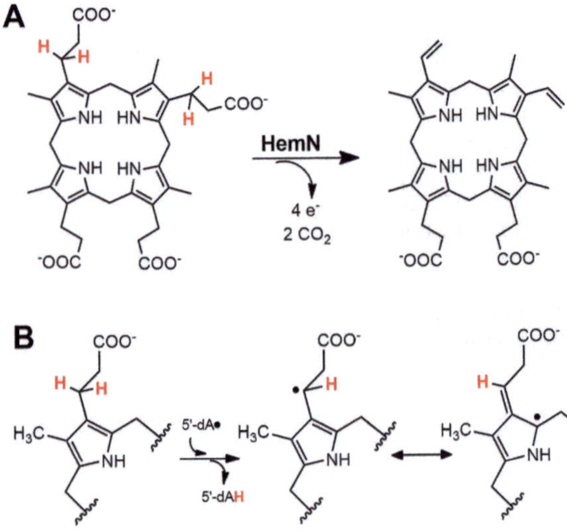

out by coproporphyrinogen III oxidase (HemN), a radical SAM enzyme that contains a single [4Fe–4S] cluster. It was postulated that the 5′-dA• formed from the reductive cleavage of SAM abstracts the *pro-S*-hydrogen atom at the β-carbon of the substrate propionate side chain, which initiates cleavage of the C_α–COO^- bond.

Therefore, an intermediate substrate radical species was expected during the catalytic reaction, and was indeed observed [85] when the substrate, coproporphyrinogen III, was added to a sample of reduced HemN along with an excess of SAM. The ability to detect this radical signal down to a temperature of 5 K suggested that the radical is in close proximity to the $[4Fe–4S]^{2+}$ cluster. To provide further information about the identity of the radical, the authors synthesized ^2H- and ^{15}N-labeled coproporphyrinogens and proved that the obtained molecules are efficient substrates for HemN. While ^{15}N-labeling did not result in any significant modification of the EPR spectra of the radical intermediate, selective deuterium labeling at different positions resulted in substantial narrowing of the EPR signal. The analysis of the spectra obtained allowed the authors to conclude that the radical species observed is a substrate-based with allylic character. The radical was also localized on the β-carbon of the carboxylate appendage and on the ring, as is evident from the effect of specific deuterium labeling of the meso-positions of coproporphyrinogen III on the EPR spectra.

Cfr/RlmN

Cfr and RlmN belong to the class A family of RS methylases. Both enzymes work on the same base, adenosine 2503 (A2503), of 23S rRNA. RlmN specifically installs a methyl group on the C2 position of the adenine ring of A2503 of 23S rRNA as

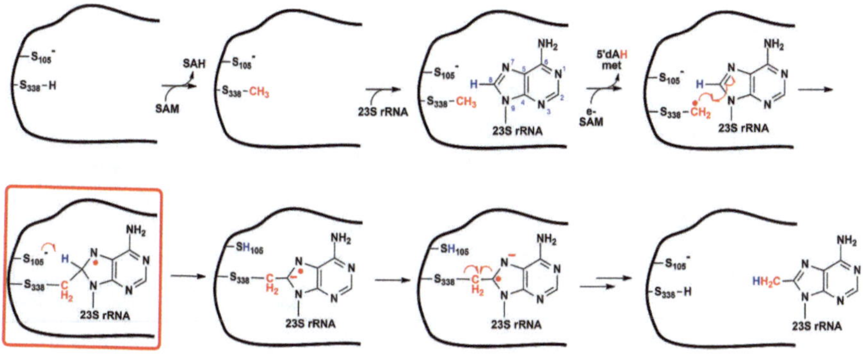

Scheme 4 Proposed radical SAM reaction mechanism of Cfr

well as on the C2 position of adenosine 37 (A37) in several tRNAs in E. *coli* and other organisms. Cfr, however, acts only on A2503 of 23S rRNA, but methylates both the C8 and the C2 carbons, with C8 methylation taking place before C2 methylation. C2 methylation of A2503 of rRNA and A37 of tRNA by RlmN plays a key role in translational fidelity and efficiency. Cfr-mediated C8 methylation, however, confers resistance to several classes of antibiotics that inhibit the peptidyl transferase activity of the bacterial ribosome, such as phenicols, lincosamides, oxazolidinones, pleuromutilins, and streptogramin A. Since 2007, cases of *cfr*-positive staphylococcal isolates from hospitalized patients in several countries, including the USA, have been reported, suggesting that this mechanism of antibiotic resistance is readily spreading, even across different staphylococcal species.

Early studies of the RlmN and Cfr reactions demonstrated that two equivalents of SAM are necessary, one to produce the 5′-dA• and a second to donate a methyl group [86]. However, the proposed mechanisms would require abstraction of hydrogen atoms from sp^2-hybridized carbon centers that exhibit C–H bond-dissociation energies above 100 kcal/mol. An alternative mechanism for the reaction was postulated upon finding that the appended methyl carbon is first appended to a conserved cysteinyl residue in the active site of each enzyme as well as the results of isotope tracer experiments that showed that the 5′-dA• abstracts a hydrogen atom from the appended methyl group [87, 88]. Scheme 4 shows the transformations proposed for Cfr. For RlmN, a similar hypothesis has been proposed.

The key element of the proposed mechanism is the addition of the resulting Cys-appended methylene radical onto C8 (Cfr) or C2 (RlmN) of A2503 to generate a protein–nucleic acid cross-linked species containing an unpaired electron. By performing this radical addition, RlmN and Cfr transiently change the hybridization of the carbon center undergoing methylation from sp^2- to sp^3-hybridized.

An EPR study performed by Grove et al. provided direct evidence for such an intermediate in the reaction of Cfr with a small, 155-mer rRNA substrate analogue [88]. When Cfr was combined with the 155-mer RNA and SAM and then rapidly mixed with dithionite to initiate turnover before loading into an EPR tube and freez-

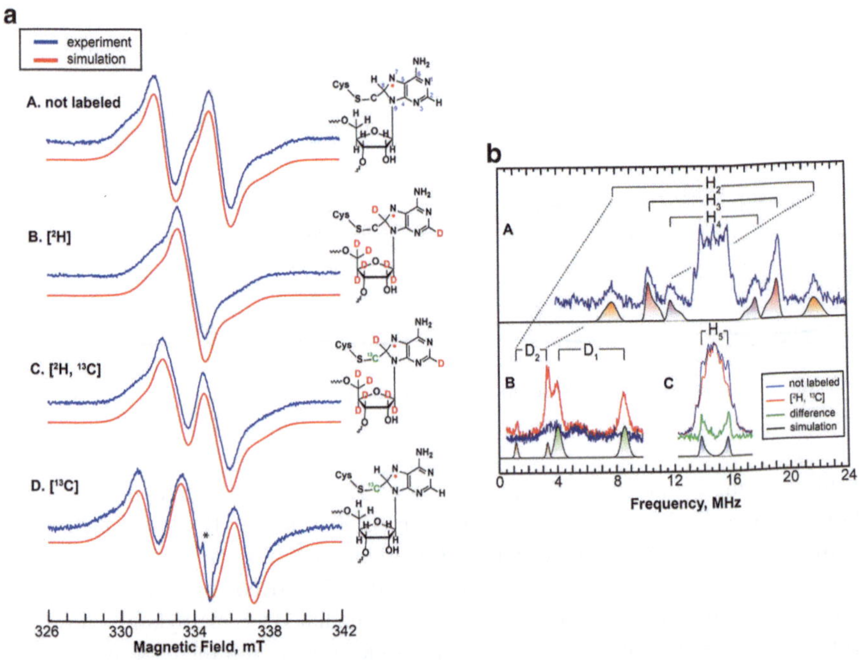

Fig. 17 X-band CW EPR (**a**) and ENDOR (**b**) spectroscopic characterization of the organic cross-linked radical intermediate in the reaction of Cfr with 155-mer RNA [88]

ing, a doublet-shaped was observed by CW EPR that had spectroscopic characteristics typical for an organic radical species. The EPR spectrum was consistent with the presence of one strong isotropic ^{1}H HF splitting and one anisotropic ^{14}N HF splitting, which coincided with the proposed electronic structure of the radical intermediate, i.e. an N7-based radical and an *sp*^{3}-hybridized C8 that retains its proton. Use of a 155-mer RNA substrate that was deuterated in the adenosine rings resulted in a collapse of the doublet into a single line, confirming that the observed species is an adenosine-based radical. To confirm a cross-linked RNA-Cfr configuration, the authors prepared Cfr such that the enzyme was unmethylated as-isolated and then supplied ^{13}C-met-SAM *in vitro* in order to install a ^{13}C label onto the mCys338 (see Fig. 17). The spectrum of the radical species obtained using such protein is dominated by clear a signature of a strong ^{13}C HF splitting when deuterated 155-mer RNA is used or adds to the ^{1}H8 splitting when unlabeled substrate is used, thus confirming the addition of the mCys to the adenosine ring.

Using a combination of Mims ENDOR and Davies ENDOR to observe ^{1}H and ^{2}H signals, five total HF coupling constants were extracted and assigned to the four protons of the adenosine ring and the C1′ proton. The obtained HF coupling constants indicated that although the spin density is significantly delocalized over the adenosine ring, the majority of the spin density is located on N7 and that C8 is *sp*^{3}-hybridized. The close resemblance of the DFT-calculated HF coupling constants to the ones obtained experimentally confirmed the overall structure of the cross-linked species.

The kinetic and chemical competence of the obtained radical species was confirmed by comparing the time dependence of the EPR signal and the product formation. The rate of product formation predicted from the time dependence of the EPR signal matched the one observed for product formation using liquid chromatography-mass spectrometry, thus providing the final proof that the radical species observed is indeed the intermediate proposed in Scheme 4.

In the case of *RlmN*, no cross-linked radical species could be observed for the wild-type protein. However, in an effort to understand role of a strictly conserved C118, Silakov et al. found that the inactive C118S and C118A variants of RlmN accumulate a substrate-derived paramagnetic species with an EPR lineshape very similar to the one observed in Cfr. Using an isotope labeling strategy similar to the one presented above permitted the assignment of the radical to a cross-linked species in which the radical is significantly delocalized over the adenine ring and an sp^3-hybridized C2 carbon. The analogous C108A Cfr variant was prepared and was found to stabilize the radical species as well. Therefore, it was concluded that the conserved residues, $C108^{Cfr}$ and $C118^{RlmN}$, are crucial for resolving the cross-linked radical species.

HydG

HydG and HydE are radical SAM enzymes that are involved in the maturation of the H-cluster active center of [Fe-Fe]-hydrogenases that catalyze the reversible heterolytic splitting of molecular hydrogen. The efficiency of the [Fe-Fe] hydrogenase as a hydrogen catalyst triggered large interest in this family of enzymes. The H-cluster is an unusual construct consisting of two subunits: a ferredoxin-like $[4Fe–4S]_H$ cluster and a unique $2Fe_H$ subcluster [89–92]. The $2Fe_H$ cluster is a dangling unit that is stabilized in the protein pocket by means of a H-bonding network, electrostatic interactions, and a single covalent link to the protein via one of the cysteines that coordinates the [4Fe–4S] "cubane". Both irons in the bi-nuclear site are coordinated by CO and CN^- ligands. Another unusual feature is the dithiolate bridge whose central atom was identified as a nitrogen by means of pulse EPR methods [93, 94] and proposed to serve as a base during heterolytic cleavage of molecular hydrogen. Using *in vitro* maturation, an exogenously added tyrosine was found to be crucial for the assembly of the cluster [95]. Isotopic labeling of tyrosine has confirmed that the backbone of this molecule is the source of the CO and CN^- ligands [96]. It was also identified that free tyrosine is the substrate for HydG, which degrades tyrosine into *p*-cresol and dehydroglycine. Free CN^- is released into the medium, where it was detected by upon being converted into a fluorescent derivative. Based on the biochemical and spectroscopic evidence, a reaction mechanism was proposed, in which abstraction of a hydrogen from the amino group by 5′-dA• triggers homolytic cleavage of the

C_α–C_β bond, resulting in the dehydroglycine intermediate that is used to generate the CO and CN^- ligands. As described above, the auxiliary cluster acts as a scaffold that eventually forms a $Fe(CN)_x(CO)_y$ unit, a building block of the $[2Fe]_H$ subcluster.

The radical species formed upon C_α–C_β scission has been investigated by EPR. Reacting wildtype HydG with L-tyrosine, SAM and a reducing agent (DTH) resulted in the transient appearance of a complex signal in the g = 2 region that peaks at about 2s. The signal was susceptible to isotope labeling of tyrosine, illustrating that the observed signal corresponds to a molecule derived from the tyrosine. When the reaction was run with 3,5-$^2H_2$-labeled Tyr the EPR lineshape was unchanged, which is consistent with the predicted C_α–C_β cleavage. Similarly, the authors noted no significant change when ^{17}O-labeled Tyr was used. On the other hand, the majority of HF splitting vanishes when β,β-$^2H_2$ tyrosine was used. Therefore, the authors determined that the majority of the spin density is localized on C_β. The fact that $^{13}C_\alpha$ labeling had no effect on the observed signal illustrated that the observed species is a 4-oxidobenzyl (4OB•) radical, therefore confirming that C_α–C_β cleavage takes place. DFT calculations also confirmed this assignment. Overall, the study convincingly illustrated the presence of an intermediate 4-oxidobenzyl radical species that is a result of a cleavage of the tyrosine (see Fig. 18).

LipA

Lipoyl synthase (LipA) catalyzes the final step in the *de novo* biosynthesis of the lipoyl cofactor—insertion of sulfur atoms at C6 and C8 of an *n*-octanoyl chain attached via an amide linkage to a target lysine residue of a lipoyl carrier protein (LCP). Recent studies have provided evidence that the auxiliary [4Fe–4S] cluster is the sacrificial donor of the requisite sulfur atoms [7]. Sequential addition of two equivalents of SAM resulted in the decay of the $[4Fe–4S]_{aux}$ first to a [3Fe–4S] cluster after the first equivalent and then to a [2Fe–2S] cluster and other mono-Fe species upon completion of the reaction. Two equivalents of the 5′-dA• are required to generate one equivalent of the lipoyl cofactor, and the two 5′-dA• have been shown to abstract H• sequentially from C6 and then C8 of octanoyl chain attached to LCPs or to short peptide substrate surrogates. These studies suggest that in the first half of the reaction, the octanoyl–LCP becomes cross-linked to LipA through the auxiliary [4Fe–4S] cluster.

To verify the proposed mechanism, Lanz et al. [98] used a two-carbon shorter substrate analogue, 2,4-hexadienoic acid, attached to the appropriate lysine residue of the *Ec*H protein of the glycine cleavage system. The authors noted that this substrate does undergoes sulfur insertion by LipA , although at a much slower rate that the native substrate. Through the course of the reaction, a radical with a complex EPR lineshape was observed (Fig. 19). Using selective ^2H-isotopic labeling, it was shown that the radical is located on the 2,4-hexadienoic acid. A total of six ^1H HF coupling constants were extracted using a combination of EPR and pulse ENDOR, indicating a substantial delocalization of the radical, with the majority of spin population located on C3 and C5 positions.

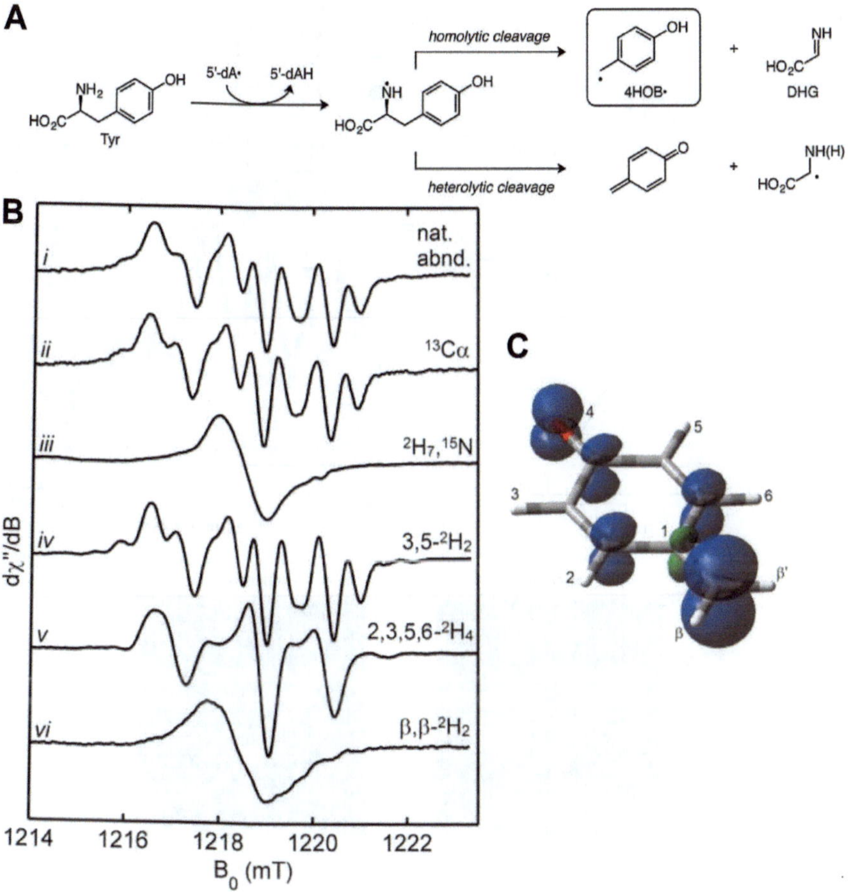

Fig. 18 Radical intermediate in the radical SAM reaction of HydG with tyrosine. (**a**) Proposed scheme of homolytic cleavage of tyrosine; (**b**) EPR spectra of the radical intermediate 4HOB• obtained using various isotopically-labeled Tyr substrates (indicated on the plot); (**c**) DFT-calculated spin distribution of the 4HOB•. Adapted with permission from Suess et al. [97]

The Q-band HYSCORE spectrum of the radical obtained using ^{57}Fe-labeled LipA showed presence of at least two ^{57}Fe signals, illustrating the proximity of the auxiliary [4Fe–4S] cluster to the substrate. The extracted HF coupling constants contained both isotropic and anisotropic components. The authors reported an effective distance extracted from the anisotropy of about 2 Å. Although this distance clearly does not represent any physical distance between atoms, it does illustrate the close proximity of the radical to the [4Fe–4S] cluster. The presence of a small isotropic component in the ^{57}Fe HF coupling was rationalized by a weak Heisenberg exchange interaction with individual iron ions.

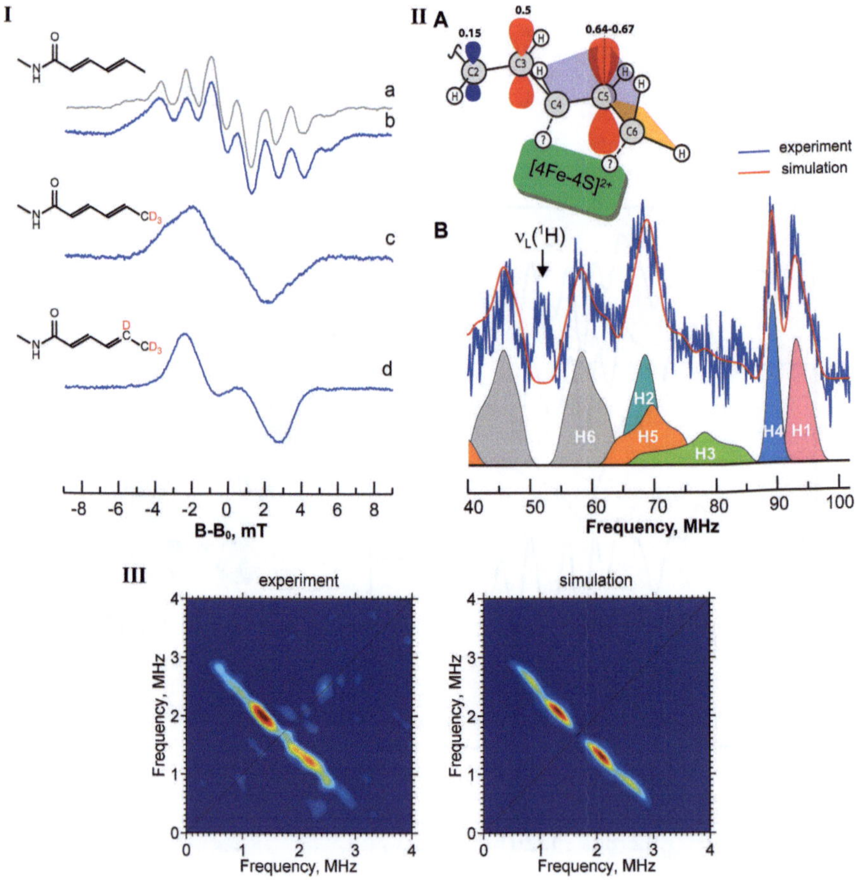

Fig. 19 X-band CW EPR spectra of the radical species of LipA obtained using unlabeled (*b*) and different ^2H isotopically labeled 2,4-hexadienoic acid (*c*, *d*) as well as a pseudomodulated Q-band FID-detected EPR spectrum of the same species (*a*) (**a**). (**b**) Q-band Davies ENDOR spectrum of the radical species (*B*) and the scheme of spin density distribution (*A*). (**c**) Q-band ^{57}Fe HYSCORE spectra of the radical species in LipA and corresponding simulation accounting for the presence of two ^{57}Fe HF couplings. Adapted from Lanz et al. [98]

Concluding Remarks

In this chapter we have presented several examples of applications of EPR spectroscopic techniques that have been used to investigate the functionality of radical SAM enzymes. This is by no means a comprehensive review of all the studies that were performed on these enzymes. In this chapter, we hoped to provide an overview of the methods that were applied so far to understand this incredibly versatile superfamily. EPR spectroscopy was used for two major functions in studying these enzymes: (1) obtain structural information about the relative arrangement of various components of the reactions and (2) resolve the identity of and obtain detailed information about the reactive intermediates.

The data accumulated to date has proven that CW EPR spectra of the $[4Fe–4S]_{RS}$ clusters are generally very sensitive to SAM binding, and one can utilize this effect as a sign of the interaction. However, we would like to note that our recent study of the B_{12}-dependent RS enzyme TsrM showed that exceptions do exist, and it is apparently possible to significantly perturb the EPR spectrum of the $[4Fe–4S]_{RS}$ cluster without bidentate SAM binding. Therefore, use of either ENDOR or HYSCORE techniques to detect ^{14}N HF interaction from the amino group is necessary to confirm the bidentate binding of SAM to the RS cluster. The data from both our studies and others suggest that the simplest way to clarify whether SAM binds is to perform a single HYSCORE experiment at the maximum absorption of the EPR signal of the $[4Fe\text{-}4S]_{RS}^{1+}$ cluster and use the position of the double quantum ridges to estimate the ^{14}N isotropic HF coupling constant. Rather interestingly, this parameter seems to vary from system to system, which is an effect that needs to be investigated systematically in conjunction with mechanistic studies.

CW EPR methods remain the most convenient technique to study organic radical intermediates, especially when combined with isotope labeling. However, as has been shown above, advanced EPR methods can significantly improve our understanding of the electronic and structural environment around the radical. Although pulse EPR methods are far more demanding in terms of technology and expertise, the details obtained are worth the effort. Moreover, the work of Hoffman and coworkers on LAM showed that a stable radical generated from a SAM analogue—*an*SAM—upon reductive cleavage can provide an important probe to observe the molecules in the vicinity of the radical, and thus provide a "t = 0" snapshot of the structural arrangement within the inner pocket. One can utilize such a method in combination with site-specific isotope labeling to decipher the site of attack for 5′-dA•, which in some cases can be as important as resolving reaction intermediates for understanding the reaction mechanism.

References

1. Pegg SC-H et al (2006) Leveraging enzyme structure–function relationships for functional inference and experimental design: the structure–function linkage database. Biochemistry 45:2545–2555
2. Akiva E et al (2014) The structure–function linkage database. Nucleic Acids Res 42:D521–D530
3. Sofia HJ, Chen G, Hetzler BG, Reyes-Spindola JF, Miller NE (2001) Radical SAM, a novel protein superfamily linking unresolved steps in familiar biosynthetic pathways with radical mechanisms: functional characterization using new analysis and information visualization methods. Nucleic Acids Res 29:1097–1106
4. Cosper NJ, Booker SJ, Ruzicka F, Frey PA, Scott RA (2000) Direct FeS cluster involvement in generation of a radical in lysine 2,3-aminomutase. Biochemistry 39:15668–15673
5. Chen DW, Walsby C, Hoffman BM, Frey PA (2003) Coordination and mechanism of reversible cleavage of S-adenosylmethionine by the [4Fe-4S] center in lysine 2,3-aminomutase. J Am Chem Soc 125:11788–11789

6. Krebs C, Broderick WE, Henshaw TF, Broderick JB, Huynh BH (2002) Coordination of adenosylmethionine to a unique iron site of the [4Fe-4S] of pyruvate formate-lyase activating enzyme: A Mossbauer spectroscopic study. J Am Chem Soc 124:912–913
7. Walsby CJ, Ortillo D, Broderick WE, Broderick JB, Hoffman BM (2002) An anchoring role for FeS clusters: chelation of the amino acid moiety of S-adenosylmethionine to the unique iron site of the [4Fe-4S] cluster of pyruvate formate-lyase activating enzyme. J Am Chem Soc 124:11270–11271
8. Walsby CJ et al (2002) Electron-nuclear double resonance spectroscopic evidence that S-adenosylmethionine binds in contact with the catalytically active [4Fe-4S](+) cluster of pyruvate formate-lyase activating enzyme. J Am Chem Soc 124:3143–3151
9. Broderick JB, Duffus BR, Duschene KS, Shepard EM (2014) Radical S-adenosylmethionine enzymes. Chem Rev 114:4229–4317
10. Frey PA, Magnusson OT (2003) S-adenosylmethionine: a wolf in sheep's clothing, or a rich man's adenosylcobalamin? Chem Rev 103:2129–2148
11. Lanz ND, Booker SJ (2012) Identification and function of auxiliary iron–sulfur clusters in radical SAM enzymes. Biochim Biophys Acta 1824:1196–1212
12. Lanz ND, Booker SJ (2015) Auxiliary iron–sulfur cofactors in radical SAM enzymes. Biochim Biophys Acta 1853:1316–1334
13. Reed GH, Poyner RR (2015) Fourier deconvolution methods for resolution enhancement in continuous-wave EPR spectroscopy. Methods Enzymol 563:23–36
14. Kauppinen JK, Moffatt DJ, Mantsch HH, Cameron DG (1981) Fourier self-deconvolution: a method for resolving intrinsically overlapped bands. Appl Spectrosc 35:271–276
15. Harmer J, Mitrikas G, Schweiger A (2009) In: Berliner L, Hanson G (eds) High resolution EPR, vol 28. Springer, New York, pp 13–61
16. Schweiger A, Jeschke G (2001) Principles of pulse electron paramagnetic resonance. Oxford University Press, Oxford
17. Petrovich R, Ruzicka F, Reed G, Frey P (1992) Characterization of iron-sulfur clusters in lysine 2,3-aminomutase by electron-paramagnetic resonance spectroscopy. Biochemistry 31:10774–10781
18. Ruszczycky MW, Choi S, Mansoorabadi SO, Liu H (2011) Mechanistic studies of the radical s-adenosyl-l-methionine enzyme desii: EPR characterization of a radical intermediate generated during its catalyzed dehydrogenation of TDP-D-quinovose. J Am Chem Soc 133:7292–7295
19. Padovani D, Thomas F, Trautwein AX, Mulliez E, Fontecave M (2001) Activation of class III ribonucleotide reductase from E-coli. The electron transfer from the iron-sulfur center to S-adenosylmethionine. Biochemistry 40:6713–6719
20. Krebs C, Henshaw TF, Cheek J, Huynh BH, Broderick JB (2000) Conversion of 3Fe-4S to 4Fe-4S clusters in native pyruvate formate-lyase activating enzyme: mossbauer characterization and implications for mechanism. J Am Chem Soc 122:12497–12506
21. Lieder KW et al (1998) S-adenosylmethionine-dependent reduction of lysine 2,3-aminomutase and observation of the catalytically functional iron-sulfur centers by electron paramagnetic resonance. Biochemistry 37:2578–2585
22. Hinckley GT, Frey PA (2006) Cofactor dependence of reduction potentials for [4Fe-4S](2+/1+) in lysine 2,3-aminomutase. Biochemistry 45:3219–3225
23. Hinckley GT, Ruzicka FJ, Thompson MJ, Blackburn GM, Frey PA (2003) Adenosyl coenzyme and pH dependence of the [4Fe-4S](2+/1+) transition in lysine 2,3-aminomutase. Arch Biochem Biophys 414:34–39
24. Hinckley GT, Frey PA (2006) An adaptable spectroelectrochemical titrator: The midpoint reduction potential of the iron–sulfur center in lysine 2,3-aminomutase. Anal Biochem 349:103–111
25. Broderick JB et al (1997) Pyruvate formate-lyase activating enzyme is an iron-sulfur protein. J Am Chem Soc 119:7396–7397
26. Maiocco SJ, Grove TL, Booker SJ, Elliott SJ (2015) Electrochemical resolution of the [4Fe-4S] centers of the AdoMet radical enzyme BtrN: evidence of proton coupling and an unusual, low-potential auxiliary cluster. J Am Chem Soc 137:8664–8667

27. Liu A, Graslund A (2000) Electron paramagnetic resonance evidence for a novel interconversion of [3Fe-4S](+) and [4Fe-4S](+) clusters with endogenous iron and sulfide in anaerobic ribonucleotide reductase activase in vitro. J Biol Chem 275:12367–12373
28. Silver SC et al (2010) Complete stereospecific repair of a synthetic dinucleotide spore photoproduct by spore photoproduct lyase. J Biol Inorg Chem 15:943–955
29. Rebeil R, Nicholson WL (2001) The subunit structure and catalytic mechanism of the Bacillus subtilis DNA repair enzyme spore photoproduct lyase. Proc Natl Acad Sci U S A 98:9038–9043
30. Yang L et al (2012) Mechanistic studies of the spore photoproduct lyase via a single cysteine mutation. Biochemistry 51:7173–7188
31. Kriek M et al (2007) Thiazole synthase from escherichia coli: an investigation of the substrates and purified proteins required for activity in vitro. J Biol Chem 282:17413–17423
32. Paraskevopoulou C, Fairhurst SA, Lowe DJ, Brick P, Onesti S (2006) The elongator subunit Elp3 contains a Fe4S4 cluster and binds S-adenosylmethionine. Mol Microbiol 59:795–806
33. McGlynn SE et al (2010) Identification and characterization of a novel member of the radical AdoMet enzyme superfamily and implications for the biosynthesis of the HMD hydrogenase active site cofactor. J Bacteriol 192:595–598
34. Brindley AA, Zajicek R, Warren MJ, Ferguson SJ, Rigby SEJ (2010) NirJ, a radical SAM family member of the d(1) heme biogenesis cluster. FEBS Lett 584:2461–2466
35. Zhang Q et al (2011) Characterization of NocL involved in thiopeptide nocathiacin i biosynthesis a [4Fe-4S] cluster and the catalysis of a radical s-adenosylmethionine enzyme. J Biol Chem 286:21287–21294
36. Yokoyama K, Ohmori D, Kudo F, Eguchi T (2008) Mechanistic study on the reaction of a radical SAM dehydrogenase BtrN by electron paramagnetic resonance spectroscopy. Biochemistry 47:8950–8960
37. Kuchenreuther JM et al (2013) A Radical intermediate in tyrosine scission to the CO and CN- ligands of FeFe hydrogenase. Science 342:472–475
38. Perche-Letuvee P et al (2012) 4-Demethylwyosine synthase from pyrococcus abyssi Is a radical-S-adenosyl-L-methionine enzyme with an additional [4Fe-4S](+2) cluster that interacts with the pyruvate co-substrate. J Biol Chem 287:41174–41185
39. McCarty RM, Krebs C, Bandarian V (2013) Spectroscopic, steady-state kinetic, and mechanistic characterization of the radical SAM enzyme QueE, which catalyzes a complex cyclization reaction in the biosynthesis of 7-deazapurines. Biochemistry 52:188–198
40. Layer G et al (2005) Radical S-adenosylmethionine enzyme coproporphyrinogen III oxidase HemN - functional features of the [4Fe-4S] cluster and the two bound S-adenosyl-L-methionines. J Biol Chem 280:29038–29046
41. Szu P-H, Ruszczycky MW, Choi S, Yan F, Liu H (2009) Characterization and mechanistic studies of desll: a radical S-Adenosyl-L-methionine enzyme involved in the biosynthesis of TDP-D-Desosamine. J Am Chem Soc 131:14030–14042
42. Kamat SS, Williams HJ, Dangott LJ, Chakrabarti M, Raushel FM (2013) The catalytic mechanism for aerobic formation of methane by bacteria. Nature 497:132–136
43. Blaszczyk AJ et al (2016) Spectroscopic and electrochemical characterization of the iron–sulfur and cobalamin cofactors of TsrM, an unusual radical S -adenosylmethionine methylase. J Am Chem Soc 138:3416–3426
44. Walsby CJ et al (2005) Spectroscopic approaches to elucidating novel iron-sulfur chemistry in the 'Radical-SAM' protein superfamily. Inorg Chem 44:727–741
45. Werst MM, Kennedy MC, Beinert H, Hoffman BM (1990) Oxygen-17, proton, and deuterium electron nuclear double resonance characterization of solvent, substrate, and inhibitor binding to the iron-sulfur $[4Fe-4S]^+$ cluster of aconitase. Biochemistry 29:10526–10532
46. Gambarelli S, Luttringer F, Padovani D, Mulliez E, Fontecave M (2005) Activation of the anaerobic ribonucleotide reductase by S-adenosylmethionine. Chembiochem 6:1960–1962
47. Tyryshkin AM, Dikanov SA, Reijerse EJ, Burgard C, Hüttermann J (1999) Characterization of bimodal coordination structure in nitrosyl heme complexes through hyperfine couplings with pyrrole and protein nitrogens. J Am Chem Soc 121:3396–3406
48. Chandor A et al (2007) Characterization of the DNA repair spore photoproduct lyase enzyme from Clostridium acetobutylicum: a radical-SAM enzyme. C R Chim 10:756–765

49. Horitani M et al (2016) Radical SAM catalysis via an organometallic intermediate with an Fe-[5'-C]-deoxyadenosyl bond. Science 352:822–825
50. Grove TL, Lee K-H, St. Clair J, Krebs C, Booker SJ (2008) In vitro characterization of AtsB, a radical SAM formylglycine-generating enzyme that contains three [4Fe-4S] clusters. Biochemistry 47:7523–7538
51. Grove TL, Ahlum JH, Sharma P, Krebs C, Booker SJ (2010) A consensus mechanism for radical SAM-dependent dehydrogenation? BtrN contains two [4Fe-4S] clusters. Biochemistry 49:3783–3785
52. Lees NS et al (2009) ENDOR spectroscopy shows that guanine N1 binds to [4Fe-4S] cluster II of the S-Adenosylmethionine-dependent enzyme MoaA: mechanistic implications. J Am Chem Soc 131:9184–9185
53. Forouhar F et al (2013) Two Fe-S clusters catalyze sulfur insertion by radical-SAM methylthiotransferases. Nat Chem Biol 9:333–338
54. Kuchenreuther JM et al (2013) Nuclear resonance vibrational spectroscopy and electron paramagnetic resonance spectroscopy of Fe-57-enriched [FeFe] hydrogenase indicate stepwise assembly of the H-cluster. Biochemistry 52:818–826
55. Kuchenreuther JM, Britt RD, Swartz JR (2012) New insights into [FeFe] hydrogenase activation and maturase function. PLoS One 7:e45850
56. Rubach JK, Brazzolotto X, Gaillard J, Fontecave M (2005) Biochemical characterization of the HydE and HydG iron-only hydrogenase maturation enzymes from Thermatoga maritima. FEBS Lett 579:5055–5060
57. Czech I, Silakov A, Lubitz W, Happe T (2010) The [FeFe]-hydrogenase maturase HydF from Clostridium acetobutylicum contains a CO and CN- ligated iron cofactor. FEBS Lett 584:638–642
58. Dinis P et al (2015) X-ray crystallographic and EPR spectroscopic analysis of HydG, a maturase in [FeFe]-hydrogenase H-cluster assembly. Proc Natl Acad Sci 112:1362–1367
59. Suess DLM et al (2015) Cysteine as a ligand platform in the biosynthesis of the FeFe hydrogenase H cluster. Proc Natl Acad Sci 112:11455–11460
60. Ugulava NB, Gibney BR, Jarrett JT (2001) Biotin synthase contains two distinct iron-sulfur cluster binding sites: chemical and spectroelectrochemical analysis of iron-sulfur cluster interconversions. Biochemistry 40:8343–8351
61. Ugulava NB, Surerus KK, Jarrett JT (2002) Evidence from Mosssbauer spectroscopy for distinct [2Fe-2S](2+) and [4Fe-4S](2+) cluster binding sites in biotin synthase from Escherichia coli. J Am Chem Soc 124:9050–9051
62. Berkovitch F, Nicolet Y, Wan JT, Jarrett JT, Drennan CL (2004) Crystal structure of biotin synthase, an S-adenosylmethionine-dependent radical enzyme. Science 303:76–79
63. Bui BTS et al (1998) Biotin synthase mechanism: on the origin of sulphur. FEBS Lett 440:226–230
64. Bui BTS, Mattioli TA, Florentin D, Bolbach G, Marquet A (2006) Escherichia coli biotin synthase produces selenobiotin. Further evidence of the involvement of the [2Fe-2S](2+) cluster in the sulfur insertion step. Biochemistry 45:3824–3834
65. Taylor AM, Stoll S, Britt RD, Jarrett JT (2011) Reduction of the [2Fe-2S] cluster accompanies formation of the intermediate 9-mercaptodethiobiotin in escherichia coli biotin synthase. Biochemistry 50:7953–7963
66. Jameson GNL, Cosper MM, Hernandez HL, Johnson MK, Huynh BH (2004) Role of the [2Fe-2S] cluster in recombinant Escherichia coli biotin synthase. Biochemistry 43:2022–2031
67. Fugate CJ et al (2012) 9-Mercaptodethiobiotin is generated as a ligand to the [2Fe-2S](+) Cluster during the reaction catalyzed by biotin synthase from escherichia coli. J Am Chem Soc 134:9042–9045
68. Knappe J, Bohnert E, Brummer W (1965) S-adenosyl-L-methionine a component of clastic dissimilation of pyruvate. Biochim Biophys Acta 107:603–605
69. Knappe J, Schmitt T (1976) Novel reaction of S-adenosyl-L-methionine correlated with activation of pyruvate formate-lyase. Biochem Biophys Res Commun 71:1110–1117

70. Knappe J, Neugebauer F, Blaschkowski H, Ganzler M (1984) Post-translational activation introduces a free-radical into pyruvate formate-lyase. Proc Natl Acad Sci U S A 81:1332–1335
71. Unkrig V, Neugebauer FA, Knappe J (1989) The free radical of pyruvate formate-lyase. Characterization by EPR spectroscopy and involvement in catalysis as studied with the substrate-analogue hypopthosphite. Eur J Biochem 184:723–728
72. Wagner A, Frey M, Neugebauer F, Schafer W, Knappe J (1992) The free-radical in pyruvate formate-lyase is located on glycine-734. Proc Natl Acad Sci U S A 89:996–1000
73. Chirpich T, Zappia V, Costilow R, Barker H (1970) Lysine 2,3-aminomutase - purification and properties of a pyridoxal phosphate and s-adenosylmethionine-activated enzyme. J Biol Chem 245:1778–1789
74. Magnusson OT, Reed GH, Frey PA (1999) Spectroscopic evidence for the participation of an allylic analogue of the 5 '-deoxyadenosyl radical in the reaction of lysine 2,3-aminomutase. J Am Chem Soc 121:9764–9765
75. Magnusson OT, Reed GH, Frey PA (2001) Characterization of an allylic analogue of the 5 '-deoxyadenosyl radical: an intermediate in the reaction of lysine 2,3-aminomutase. Biochemistry 40:7773–7782
76. Horitani M et al (2015) Why nature uses radical SAM enzymes so widely: electron nuclear double resonance studies of lysine 2,3-aminomutase show the 5′-dAdo• 'free radical' is never free. J Am Chem Soc 137:7111–7121
77. Miller J, Bandarian V, Reed GH, Frey PA (2001) Inhibition of lysine 2,3-aminomutase by the alternative substrate 4-thialysine and characterization of the 4-thialysyl radical intermediate. Arch Biochem Biophys 387:281–288
78. Wu WM et al (2000) Lysine 2,3-aminomutase and trans-4,5-dehydrolysine: Characterization of an allylic analogue of a substrate-based radical in the catalytic mechanism? Biochemistry 39:9561–9570
79. Reed GH, Ballinger MD (1995) [24] Characterization of a radical intermediate in the lysine 2,3-aminomutase reaction. Methods Enzymol 258:362–379
80. Lees NS et al (2006) How an enzyme tames reactive intermediates: positioning of the active-site components of lysine 2,3-aminomutase during enzymatic turnover as determined by ENDOR spectroscopy. J Am Chem Soc 128:10145–10154
81. Ballinger M, Reed G, Frey P (1992) An organic radical in the lysine 2,3-aminomutase reaction. Biochemistry 31:949–953
82. Chang CH, Ballinger MD, Reed GH, Frey PA (1996) Lysine 2,3-aminomutase: rapid mix-freeze-quench electron paramagnetic resonance studies establishing the kinetic competence of a substrate-based radical intermediate. Biochemistry 35:11081–11084
83. Ballinger M, Frey P, Reed G (1992) Structure of a substrate radical intermediate in the reaction of lysine 2,3-aminomutase. Biochemistry 31:10782–10789
84. Ballinger M, Frey P, Reed G, Lobrutto R (1995) Pulsed electron-paramagnetic-resonance studies of the lysine 2,3-aminomutase substrate radical - evidence for participation of pyridoxal 5'-phosphate in a radical rearrangement. Biochemistry 34:10086–10093
85. Layer G et al (2006) The substrate radical of Escherichia coli oxygen-independent coproporphyrinogen III oxidase HemN. J Biol Chem 281:15727–15734
86. Yan F et al (2010) RImN and Cfr are Radical SAM Enzymes Involved in Methylation of Ribosomal RNA. J Am Chem Soc 132:3953–3964
87. Grove TL et al (2011) A radically different mechanism for s-adenosylmethionine-dependent methyltransferases. Science 332:604–607
88. Grove TL et al (2013) A substrate radical intermediate in catalysis by the antibiotic resistance protein Cfr. Nat Chem Biol 9:422–427
89. Peters JW, Lanzilotta WN, Lemon BJ, Seefeldt LC (1998) X-ray crystal structure of the Fe-only hydrogenase (Cpl) from Clostridium pasteurianum to 1.8 angstrom resolution. Science 282:1853–1858
90. Lemon BJ, Peters JW (1999) Binding of exogenously added carbon monoxide at the active site of the iron-only hydrogenase (CpI) from Clostridium pasteurianum. Biochemistry 38:12969–12973

91. Nicolet Y et al (2001) Crystallographic and FTIR spectroscopic evidence of changes in Fe coordination upon reduction of the active site of the Fe-only hydrogenase from Desulfovibrio desulfuricans. J Am Chem Soc 123:1596–1601
92. Nicolet Y, Piras C, Legrand P, Hatchikian CE, Fontecilla-Camps JC (1999) Desulfovibrio desulfuricans iron hydrogenase: the structure shows unusual coordination to an active site Fe binuclear center. Struct Fold Des 7:13–23
93. Silakov A, Wenk B, Reijerse E, Lubitz W (2009) N-14 HYSCORE investigation of the H-cluster of [FeFe] hydrogenase: evidence for a nitrogen in the dithiol bridge. Phys Chem Chem Phys 11:6592–6599
94. Adamska-Venkatesh A et al (2015) Spectroscopic characterization of the bridging amine in the active site of [FeFe] hydrogenase using isotopologues of the H-cluster. J Am Chem Soc 137:12744–12747
95. Kuchenreuther JM, Stapleton JA, Swartz JR (2009) Tyrosine, cysteine, and S-adenosyl methionine stimulate in vitro [FeFe] hydrogenase activation. PLoS One 4:e7565
96. Kuchenreuther JM, George SJ, Grady-Smith CS, Cramer SP, Swartz JR (2011) Cell-free H-cluster synthesis and [FeFe] hydrogenase activation: All Five CO and CN- ligands derive from tyrosine. PLoS One 6:e20346
97. Suess DLM, Kuchenreuther JM, De La Paz L, Swartz JR, Britt RD (2016) Biosynthesis of the [FeFe] hydrogenase H Cluster: a central role for the radical SAM enzyme HydG. Inorg Chem 55:478–487
98. Lanz ND et al (2015) Characterization of a radical intermediate in lipoyl cofactor biosynthesis. J Am Chem Soc 137:13216–13219

Index

G. Hanson, L. Berliner (eds.), *Future Directions in Metalloprotein and Metalloenzyme Research*, Biological Magnetic Resonance 33,
DOI 10.1007/978-3-319-59100-1

MIX
Papier aus verantwortungsvollen Quellen
Paper from responsible sources
FSC® C105338

If you have any concerns about our products,
you can contact us on
ProductSafety@springernature.com

In case Publisher is established outside the EU,
the EU authorized representative is:
Springer Nature Customer Service Center GmbH
Europaplatz 3, 69115 Heidelberg, Germany

Printed by Libri Plureos GmbH
in Hamburg, Germany